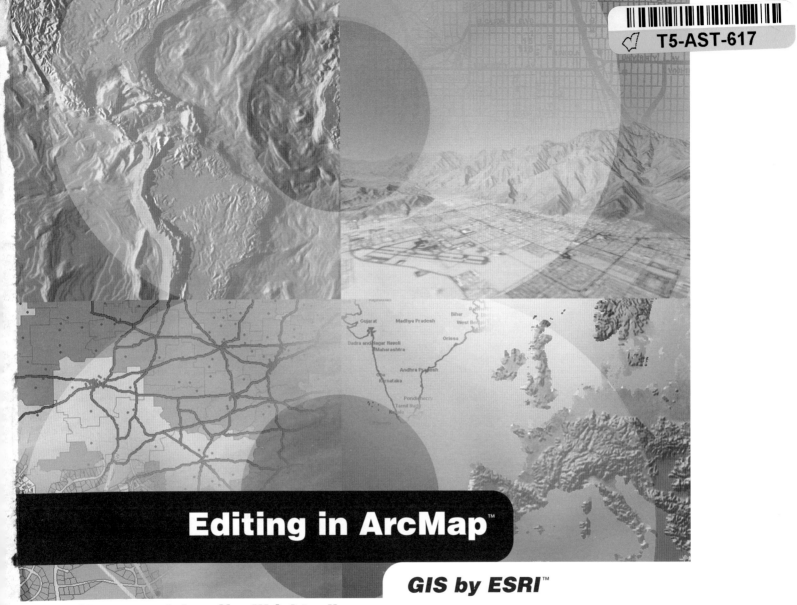

Editing in ArcMap™

GIS by ESRI™

Jeff Shaner and Jennifer Wrightsell

DATA CREDITS

Graphical Editing Map: Wilson, North Carolina

Universal Data Editor Map, Editing in data view and layout view map: Greeley Colorado
Context menus and shortcut keys map: P.F.R.A., Regina Saskatchewan Canada
Quick-Start Tutorial Data: Wilson, North Carolina, Greeley Colorado

CONTRIBUTING WRITERS

Bill Moreland, Doug Morgenthaler, Jan van Roessel, Jeff Jackson, Kristin Clark, Michelle Sakala, Robin Floyd, Steve Van Esch, Tim Hodson, Wayne Hewitt

U.S. GOVERNMENT RESTRICTED/LIMITED RIGHTS

Contents

Introduction

1

In addition to mapmaking and map-based analysis, ESRI®ArcMap™ is the application for creating and editing geographic data as well as tabular data. With ArcMap, you can edit shapefiles, coverages, and geodatabases all with one common user interface. ArcMap contains sophisticated, CAD-based editing tools that help you construct features quickly and easily while maintaining the spatial integrity of your GIS database.

Whether you use ArcView® GIS or ArcInfo™, you use the same editing tools in ArcMap to work on your geographic data. In cases where your organization has multiple users simultaneously editing on a shared geodatabase, ArcMap, in concert with ArcSDE™, provides the tools necessary to manage long editing transactions, as well as to manage versions and resolve potential conflicts.

Whether you use ArcView GIS, ArcEditor™, or ArcInfo, the goal of this book is to help you learn and use the editing capabilities in ArcMap for any level of geographic database maintenance. The next few pages highlight some of the features you will find invaluable while editing in ArcMap.

Rich suite of graphical editing tools

ArcMap helps you create and edit geographic features quickly and easily by including many of the graphic editing functions popular with the latest computer-aided design (CAD) editing packages.

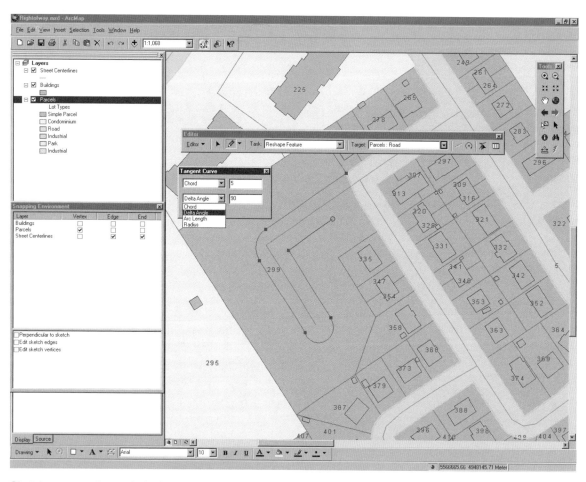

Sketch construction tools in ArcMap will allow you to quickly and accurately edit street rights-of-way.

Universal data editor

ArcMap lets you edit shapefiles, coverages, and geodatabases in their native data formats. Also, you can edit an entire folder of data at once.

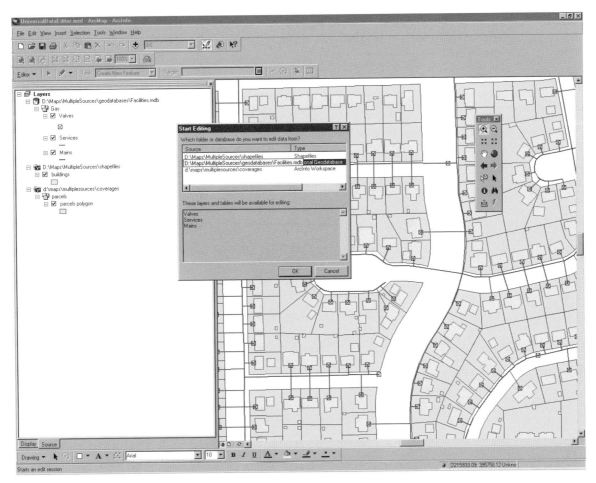

Pick the folder of data that you want to edit when you start editing in ArcMap.

Editing in data view and layout view

ArcMap provides two different ways to view a map: data view and layout view. Each view lets you look at and interact with the map in a different way. Data view hides all of the map elements on the layout such as titles, North arrows, and scale bars. In layout view, you'll see a virtual page upon which you can place and arrange map elements. You can edit your geographic data in either data view or layout view.

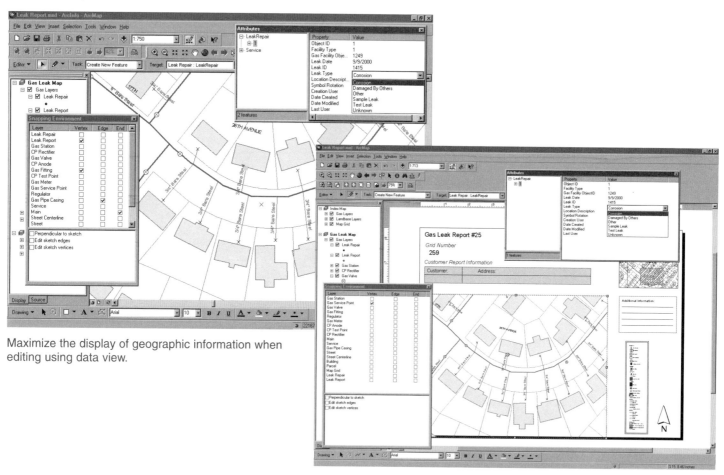

Maximize the display of geographic information when editing using data view.

When you are preparing a map, you can edit features directly using layout view.

Context menus and shortcut keys for increased productivity

ArcMap contains numerous context menus and shortcut keys to help you create and edit features quickly.

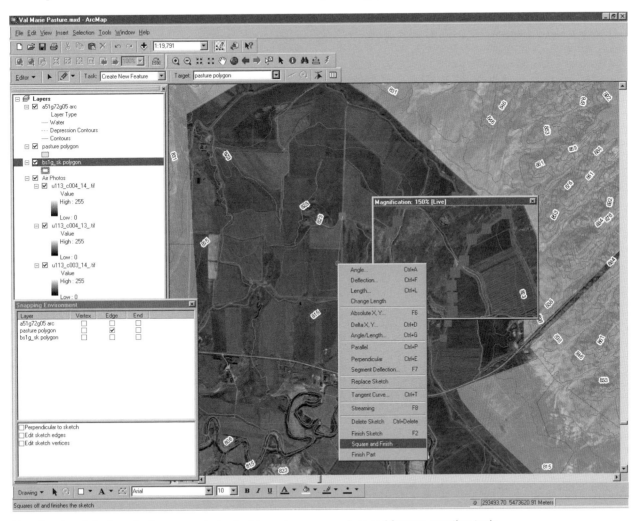

Use the sketch tool context menu and shortcut keys to access advanced feature creation tools.

Multiuser editing with version management and conflict detection

If you have several users that need to edit the same data at the same time, ArcMap can help you manage versions of your ArcSDE geodatabase.

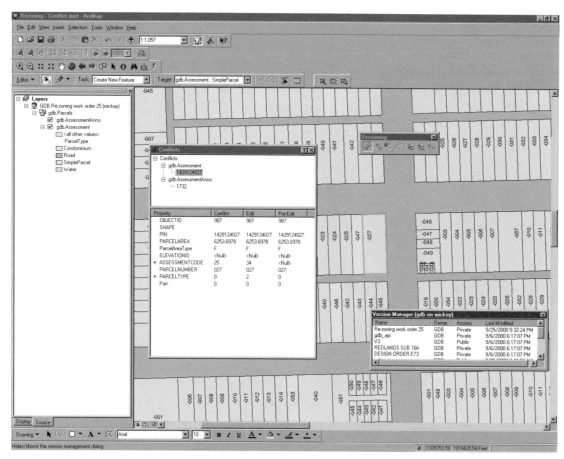

Sophisticated version management tools in ArcMap will help you maintain a multiuser editing environment.

Editing in projected space

If you've collected data from a variety of sources, chances are that not all layers contain the same coordinate system information. Using ArcMap, you can set the coordinate system for a data frame. As you add layers to your map, they are automatically transformed to that projection. That means that you can edit the shapes and attributes of a layer regardless of the coordinate system it was stored in.

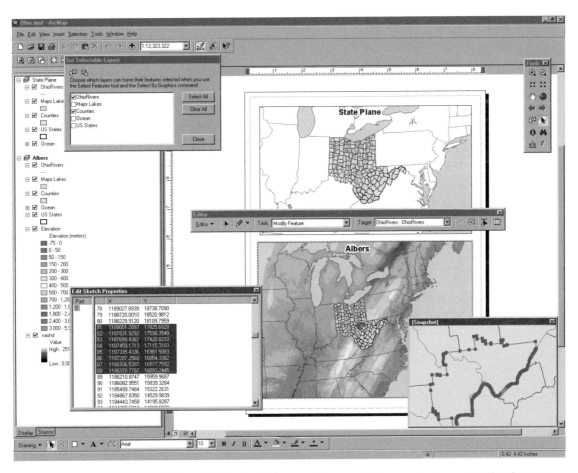

ArcMap has "project-on-the-fly" capabilities that let you edit layers in the coordinate system that is most important to you without having to transform your data.

Tips on learning how to edit in ArcMap

If you're new to GIS, remember that you don't have to learn everything about editing in ArcMap to get immediate results. Begin learning how to edit by reading Chapter 2, 'Quick-start tutorial'. In Chapter 2, you'll learn how easy it is to use the various editing tools in conjunction with ArcMap, and you'll gain insight into the steps you can take to complete certain tasks. ArcMap comes with the data used in the tutorial, so you can follow along step by step at your computer. You can also read the tutorial without using your computer.

Finding answers to your questions

Like most people, your goal is to complete your tasks while investing a minimum amount of time and effort in learning how to use software. You want intuitive, easy-to-use software that gives you immediate results, without having to read pages of documentation. However, when you do have a question, you want the answer quickly so that you can complete your task. That's what this book is all about—getting you the answers you need, when you need them.

This book describes editing tasks—from basic to advanced—that you'll perform with ArcMap. Although you can read this book from start to finish, you'll likely use it more as a reference. When you want to know how to do a particular task, such as creating a new feature, just look it up in the table of contents or index. What you'll find is a concise, step-by-step description of how to complete the task. Some chapters also include detailed information that you can read if you want to learn more about the concepts behind the tasks. You may also refer to the glossary in this book if you come across any unfamiliar GIS terms or need to refresh your memory.

About this book

This book is designed to introduce editing in ArcMap and its capabilities. The topics covered in the various tasks and the tutorial in Chapter 2 assume you are familiar with building a GIS database and the fundamentals of GIS. If you have never used a GIS before or feel you need to refresh your knowledge, please take some time to read *Getting Started with ArcGIS*, which you received in your ArcGIS package. It is not necessary to do so to continue with this book, but you should use it as a reference if you encounter tasks with which you are unfamiliar.

Chapter 3 is an introduction to editing in ArcMap and describes the basic tasks you need to know before you can start creating and editing spatial data. It's easy to create new features in ArcMap, and Chapter 4 describes the different ways you can create them.

Chapter 5 illustrates how you can connect a digitizing tablet to your computer and digitize GIS features from paper maps. When editing spatial data, often new features are created using the shapes of other features. Chapter 6 shows you how easy it is to perform these tasks in ArcMap.

Chapter 7 shows how to modify the shapes of features that already exist in your database. If you need to maintain topology when you edit features, Chapter 8 will show you how. Chapter 9 will show you how to edit the attributes of a single feature or all of the selected features in a layer.

You will find that the basic patterns of editing geographic data are the same across different data models when using the editing tools in ArcMap. However, if you are editing coverage features some tools may behave differently depending on the type of feature. Appendix A will explain that behavior in detail.

Getting help on your computer

In addition to this book, ArcMap software's online Help system is a valuable resource for learning how to use the software.

Contacting ESRI

If you need to contact ESRI for technical support, see the product registration and support card you received with ArcMap or see the topic 'Contacting Technical Support' in the 'Getting more help' book of the ArcGIS Desktop Help system. You can also visit ESRI on the Web at www.esri.com and www.arconline.esri.com for more information about ArcMap.

Quick-start tutorial

2

ArcMap has the tools you need to create and edit your spatial databases. It is easy to learn and ensures a simple, quick, and natural transition from viewing geography to editing geography.

The easiest way to learn how to edit in ArcMap is to complete the exercises in this tutorial.

Exercises 1 and 2 introduce the edit sketch, sketch tools, and edit tasks and shows you how to use them to create new features quickly and easily.

Exercise 3 walks you through the process of converting features on a paper map directly into your database using a digitizing tablet.

Exercise 4 teaches you how to move, rotate, scale, extend, trim, and modify the vertices of existing features.

Exercise 5 shows you how to create and maintain the shared boundaries between features and layers.

Exercise 6 demonstrates how you can integrate layers from computer-aided design (CAD) drawings into your database.

Each of these exercises takes between 15 and 20 minutes to complete. You have the option of working through the entire tutorial or completing each exercise one at a time.

Exercise 1: Creating polygon features

The editing tools in ArcMap make it very easy to create new features. You use edit tasks, the edit sketch, sketch tools, and snapping to create new features in ArcMap.

In this exercise, you will digitize a new polygon feature into a shapefile layer that outlines a land use study region. The study area polygon that you create needs to snap to an index grid layer that subdivides the entire geographic region. You will begin by starting ArcMap and loading a map document that contains the shapefile layer and a geodatabase that contains the index grid for the region.

Starting ArcMap and beginning editing

Before you can complete the tasks in this tutorial, you must start ArcMap and load the tutorial data.

1. Double-click a shortcut installed on your desktop or use the Programs list in your Start menu to start ArcMap.

2. Click the Open button on the Standard toolbar. Navigate to the CreatingNewFeatures.mxd map document in the Editor directory where you installed the tutorial data (C:\ArcGIS\ArcTutor is the default location).

3. Click the Editor Toolbar button on the Standard toolbar to add the Editor toolbar to ArcMap.

4. Click the Editor menu and click Start Editing.

If you only have one workspace in your map, you can start editing the map layers at this point. In this exercise, two workspaces are loaded in the map, so you will need to choose the workspace you want to edit.

5. Click the Editor folder workspace to start editing the studyarea.shp shapefile. Click OK. You will edit the geodatabase in the next exercise.

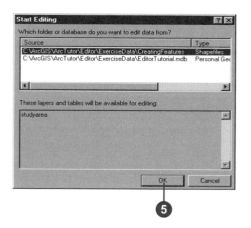

Creating a new polygon feature

This exercise focuses on creating a new study area polygon that encompasses a parcel CAD drawing. The extent of the study area is defined by the index grid lines located in an existing database. The index grid represents logical divisions within the data.

To create the new polygon, you must do "heads-up" digitizing against the index grid and snap the vertices of your new polygon to the vertices of the grid lines.

Setting the snapping environment

Before you start editing the study area shapefile, you need to set your snapping environment so that each point you add snaps to the vertices of features in the index grid. For more information about snapping, see 'Using the snapping environment' in Chapter 4.

1. Click the Editor menu and click Snapping to display the Snapping Environment window.

2. Check the Vertex check box next to the IndexGrid layer to snap the sketch vertices to the vertices of the index grid.

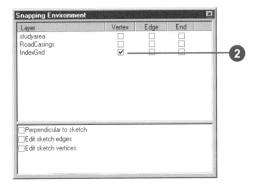

With the snapping environment set, you can create a new study area polygon. Make sure you snap each point to the thick index grid lines shown below.

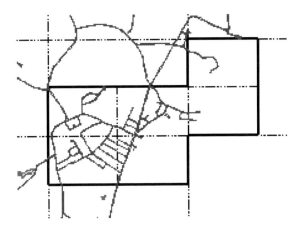

Setting the current task

Before you start digitizing a new feature, you must set the current editing task to Create New Feature.

1. Click the Current Task dropdown arrow and click Create New Feature.

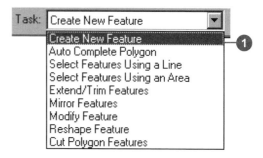

To create a new feature, you also need a target layer. The target layer determines the type of feature you will create and what layer it will be stored in. Since there is only one shapefile in the folder that you started to edit, the target layer is set to the study area shapefile by default.

Using the Sketch tool

To create a new feature using the Create New Feature task, you must first create an edit sketch. An edit sketch is a shape that you draw by digitizing vertices using the sketch construction tools located on the tool palette.

Several tools can add vertices to the sketch. You will use the Sketch tool to add the study area polygon.

1. Click the tool palette dropdown arrow and click the Sketch tool.

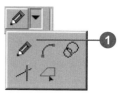

2. Click to add the first vertex of the sketch to the lower-left corner of the thick index grid lines. The vertex should snap in place.

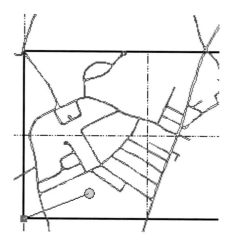

3. Click to add the remaining vertices, snapping each vertex to a corner in the index grid. Create vertices counterclockwise until you return to the point located directly above the first vertex that you placed.

Finishing the sketch

1. Press the F2 key or right-click and click Finish Sketch.

   This action adds the final sketch segment and creates the new feature.

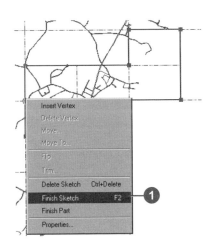

Your new study area polygon is now created. If you snapped each sketch vertex properly, the new polygon should look like the shaded polygon below.

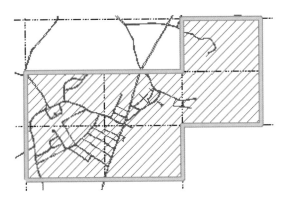

Adding attributes

The new feature you created does not contain any attribute information. Because other polygon features are present in this shapefile, distinguish your new polygon from the others by adding descriptive information about it.

You can add descriptive information for a selected feature using the Attributes dialog box or "property inspector".

1. Click the Attributes button on the Editor toolbar to add a description attribute to the new study area polygon.

2. Click the layer field for the selected feature and type "StudyArea" as a description of the feature.

Saving your edits

After you have created the new study area polygon, you can choose to save or discard your edits by stopping the edit session.

1. Click the Editor menu and click Stop Editing.

2. Click Yes to save the new study area polygon into the study area shapefile you were editing or No to discard your edits.

In this exercise you learned how to quickly and accurately create a new polygon feature. You used the Sketch tool to digitize a polygon shape while snapping each vertex to an existing vertex in another layer.

There are several other ways that you can construct new features in your GIS database. The next exercise will show you some of the more advanced methods of constructing vertices in the edit sketch.

For more detailed information about editing tasks and creating polygon features, see Chapter 4, 'Creating new features'.

Exercise 2: Creating line features

In this exercise, you will update the existing road network in your database with a new road casing line.

In building the line feature, you will learn how to use some of the more advanced construction methods offered with the Sketch tool context menu.

Editing the geodatabase

Because the road feature class exists inside a different workspace than the study area shapefile, you need to start editing the database before you can create the new line.

1. Click the Editor menu and click Start Editing. Select the personal geodatabase as the workspace that you want to edit and click OK.

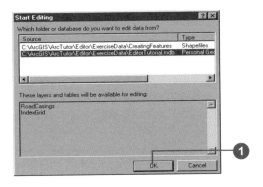

Locating the update area

Spatial bookmarks are named extents that can be saved in map documents. Creating a bookmark for areas that you visit frequently will save you time. For information on how to create and manage spatial bookmarks, see Chapter 3, 'ArcMap basics', in *Using ArcMap*.

You will now zoom to a spatial bookmark created for this exercise.

1. Click the View menu, point to Bookmarks, then click Update road casings to set the current view to the edit area of this exercise.

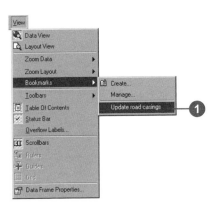

When the display refreshes, note that the top line of this road casing is missing from the layer. You must update the road casing by adding the missing line.

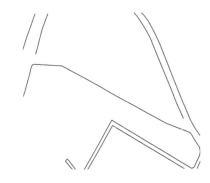

Setting the snapping environment

The endpoints of the road casing feature need to snap to adjacent casings to ensure that the new feature is connected to the existing casing features. Snapping to the end of road casing lines will help you do this.

1. Click the Editor menu and click Snapping. Check the End option for the RoadCasings layer to set snapping to the endpoint of casing features.

Digitizing

After setting the snapping environment, make sure that the target layer is set to the RoadCasings layer, and then you can start digitizing.

1. Click the tool palette dropdown arrow and click the Sketch tool.

2. Move the pointer to the broken section of the road casing in the top-left corner of the canvas. Once the pointer is inside the snapping tolerance, the snapping location (blue dot) will move away from the pointer. Click the left mouse button to add the first vertex.

Beginning construction

With the first vertex of the new road casing properly placed, you can construct the casing line feature. Your new feature will be connected to that casing.

Setting length and angle measurements

Before creating the second vertex, you must first set the length of the line.

1. Right-click the map and click Length.

2. Type a value of 15 map units and press Enter.

If you move the pointer now, notice that you can't stretch the line further than your length measurement. This is called a *constraint*. To learn more about sketch constraints, see Chapter 4, 'Creating new features'.

You must also set an angle constraint to create the second vertex.

3. Press Ctrl + A and type a value of 260 degrees. Press Enter.

Creating a curve tangent to the last segment

Next, add a curve that is tangent to the last segment you just added to the sketch. The curve will form the corner of the road casing.

1. Right-click and click Tangent curve to enter the curve information required to place the next vertex.

2. Click the first dropdown arrow and click Chord. Type "20" to set the chord length. Click the second dropdown arrow and click Delta Angle. Type "90" in the second text box for the angle measurement. Click Left to indicate that the new curve will be tangent to the left of the previous segment. Press Enter to create the curve.

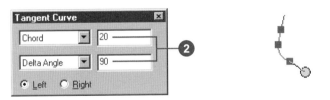

Creating a vertex relative to the last vertex

Often, construction points are calculated relative to the last point recorded. Using the Delta X, Y sketch constructor, you can add relative vertices.

1. Press Ctrl + D. Type "88" for the x-value and "-9" for the y-value. Press Enter to add the point.

Creating a vertex parallel to an existing line

You can define the angle measurement for points added to the sketch in several ways. You can set an absolute value like you did in the first step of this exercise, or you can use the angles of existing features. Quite often, road casings are constructed using the angles of road centerlines. Since we already have one road casing, we can use its angle in constructing the next segment.

1. Right-click on the lower road casing line. Click Parallel. Press Ctrl + L, type a value of 415, then press Enter.

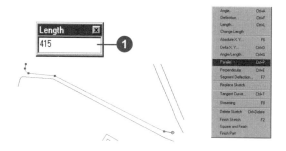

Creating a new vertex using absolute coordinates

Quite often exact x- and y-coordinate information is available for the construction of vertices. Add the next vertex by typing exact coordinates using the Absolute X, Y constructor.

1. Right-click on map and click Absolte X, Y. Type "1227820.6" in the x field, press the Tab key, and type "181460.6" in the y field. Press Enter to add the point.

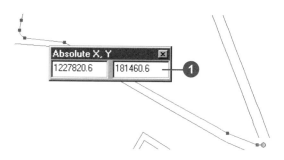

Creating a tangent curve

One final tangent curve needs to be added to the sketch before you can connect it to the existing casing and add the feature.

1. Press Ctrl + T. Type a chord length of 12 and a delta angle of 120, then press Enter to create the final curve segment.

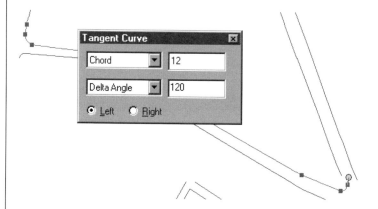

Finishing the sketch

To finish the sketch and create the feature so that it is connected to the existing casing, you need to snap the last point of the sketch to the endpoint of the existing road casing.

1. Move the pointer to the endpoint of the existing road casing until it snaps. Double-click to add the last point and create the feature.

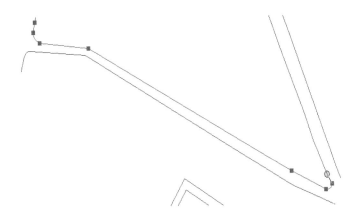

With construction now complete, you can continue to search the layer to find additional broken lines and connect them together, experimenting with these and other sketch tools and construction techniques.

The next exercise will show you how you can use the construction methods demonstrated in this exercise to capture features from a paper map directly into your GIS layers using a digitizing tablet.

Exercise 3: Using a digitizing tablet

The first step in the first exercise showed you how to "heads-up" digitize features by snapping to an existing vector source. However, often that source information is in paper form. ArcMap lets you trace over the features you are interested in capturing using a digitizing tablet connected to your computer. By digitizing data using a tablet, you can get features from almost any paper map into your GIS database.

Setting up your digitizing tablet

Before you can start digitizing, you must set up your tablet and prepare the map from which you want to digitize.

Before you can continue with this exercise, you must have properly installed the WinTab driver for your tablet and configured the buttons on your digitizing puck. If you installed the WinTab driver after installing ArcInfo, you will need to register the digitizer.dll in order to continue. For more information on digitizers, see Chapter 5, 'Using a digitizer'.

Preparing the map

You will now print the paper map from which you want to digitize and attach it to your tablet.

1. Print the "DigitizingFeatures.tif" image located in the Editor tutorial directory (where you installed the tutorial data) and tape it to your tablet using masking tape.

2. Start ArcMap if you haven't already done so. Open the DigitizingFeatures.mxd map document to register the paper map to your map document.

Registering your map for the first time

You must always register your paper map before you can begin digitizing from it. This involves establishing control points to register the paper map to the geographic space of your GIS data.

You can add control points as x,y coordinates stored in a comma-delimited American Standard Code for Information Interchange (ASCII) file and load those points into your map document when you are ready to digitize. However, for this exercise, you will establish these coordinates manually as you register the points to known map coordinates located on the paper map you printed out.

1. Click the Editor menu and click Start Editing. Click the Editor menu and click Options.

2. Click the Digitizer tab. You will create and store control points here.

The control points you add will be saved with the map document.

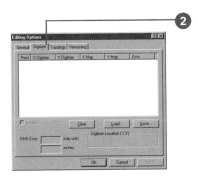

3. Starting with the upper-left corner of your paper map and working clockwise, locate the point marked "1" and click on it using the digitizer puck. Then type the associated ground x,y coordinates indicated on the paper map into the dialog next to the point entered.

4. Digitize and type values for the other three points, then examine the total root mean square (RMS) error calculation for all points. Error values are displayed in map and digitizer units. To maintain highly accurate data, the RMS error should be less than 0.004 digitizer units.

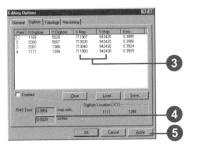

To reduce the total RMS error, you can replace points by clicking on the point you want to replace in the Editing Options menu, then clicking on the paper map to capture a new point. By reentering points with a high error value, you can reduce the total error.

5. Click Apply to accept the registration after you have reached an acceptable RMS error.

Digitizing modes

You need to enable digitizing mode once you have registered your map. Enabling digitizing mode maps the location of the puck on the tablet to a specific location on the screen.

1. Click the Enabled check box on the Digitizer tab of the Editing Options dialog box to enable digitizing mode.

2. Click OK.

Digitizing new features

You are now ready to begin digitizing new features. You will add new lot lines representing a new parcel subdivision into an existing shapefile of lot lines.

Setting the current task and target layer

Creating new features using a digitizer puck is identical to creating new features using the mouse. You must set the current task and target layer before you start digitizing.

1. Click the Current Task dropdown arrow and click Create New Feature.

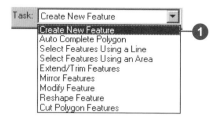

2. Click the Target layer dropdown arrow and click Lotlines to set the target layer.

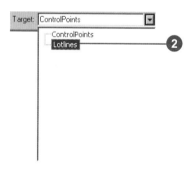

Creating new features

There are two ways to digitize features on a paper map: point mode digitizing and stream mode digitizing (streaming). You can easily toggle back and forth between point and stream mode by pressing F8.

Digitizing in point mode

Point mode is the default and most common method of digitizing paper features. To digitize features in point mode, you click to add each point or vertex. If the paper feature was drafted with considerable accuracy, you should use point mode digitizing and snapping to retain that accuracy.

1. Click the Editor menu and click Snapping. Check the Edge option for the Lotlines layer to make sure the features you digitize snap to existing edges.

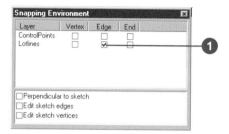

2. Trace over the frontage lot lines (lines in red) that define the road leading into the new subdivision. For straight segments, you should add a vertex where lot lines intersect. Click the Sketch tool and then click on the paper map to start adding vertices.

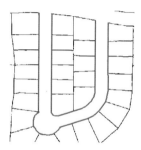

Digitizing in stream mode

When tracing line or polygon features, you may want to add vertices as you move the mouse rather than click each time you want to add a vertex. Stream mode digitizing lets you do this.

Before starting to digitize in stream mode, you need to set a stream tolerance—the interval at which the sketch adds vertices along the feature you are digitizing. The default tolerance value is 0 map units, so if you don't enter a tolerance value you may find vertices that join or overlap each other. You will also set the group tolerance for stream mode to the same value; this will let you remove numerous points in a single undo operation.

1. Click the Editor menu and click Options. Click the General tab and type a new stream tolerance value of 45 map units; make sure that the group tolerance is set to 45 as well. Click OK when you are finished setting the tolerance values.

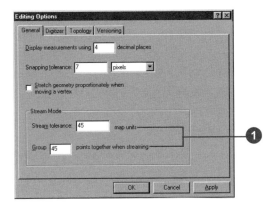

2. Snap the first point to the exterior of the new lot lines (upper-left corner), then press F8 to start digitizing in stream mode. Trace along the boundary of the lots (lines in red) until you reach the last lot (upper-right). Notice that vertices are added at consistent intervals—45 map units apart. Press F8 to stop digitizing in stream mode, snap the last vertex to the existing lot line, and double-click to finish the sketch.

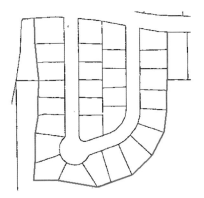

3. With the exterior lot lines digitized, proceed to digitize each remaining paper line feature required to define the lots.

   Once you are finished digitizing, you should disable the digitizer puck.

4. Click the Editor menu and click Options. Click the Digitizer tab and uncheck Enabled to disable the digitizer. Click OK.

Finishing your digitizing session

Once you have finished tracing lot lines and have disabled the digitizer puck, you can stop editing and complete the exercise by saving your edits.

1. Click the Editor menu and click Stop Editing.

2. Click Yes to save your edits.

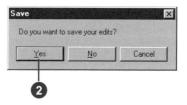

In this exercise you learned how to create new features and append your GIS database by digitizing shapes directly from a digitizing tablet. The next exercise will show you how to copy shapes from existing vector sources (CAD drawing layers) and paste them into your GIS database.

To learn more about digitizing, see Chapter 5, 'Using a digitizer'. If you need to find out if ArcMap supports your digitizing tablet, please consult the ESRI Web site (www.esri.com) for the most recent information.

Exercise 4: Editing features

In the first three exercises, you learned how to create new features in ArcMap. In this exercise, you'll learn how to copy and paste, move, rotate, scale, and extend existing features.

Opening the exercise document and starting editing

1. Start ArcMap.

2. Click the Open button on the Standard toolbar. Navigate to the EditingFeatures.mxd map document located in the Editor directory where you installed the tutorial data (C:\ArcGIS\ArcTutor is the default location).

3. Click the Editor menu and click Start Editing.

Copying and pasting features

When creating vector features of the same type as existing ones, it is more efficient to copy their shapes than to digitize over the top of them. You can copy the shapes of any vector feature that you can select in ArcMap. In this step, you will select buildings from a CAD drawing and paste them into a geodatabase layer of buildings.

1. Click the Edit tool on the Editor toolbar and drag a box around all of the new building features to select them.

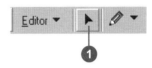

All selected CAD features should be highlighted as shown below.

2. Click the Copy button on the Standard toolbar to copy the selected features to the clipboard.

3. Set the Buildings layer as the target layer so that you can paste the copied features into it.

4. Click Paste to copy the selected building features into the target layer. The progress bar will update as each feature is copied into the target layer.

It is important to note that only the shapes are copied from the CAD file into the geodatabase. If you need to paste the attributes as well, you must use the object loader. Exercise 6 of this tutorial shows you how to do this.

Rotating features

Now that you've copied the building features into the Buildings layer of your geodatabase, you need to orient the features to fit the parcel subdivision into which you'll move them.

1. To avoid selecting features from the CAD layer (called New Buildings), uncheck it in the table of contents to hide its features.

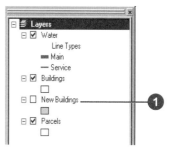

2. Click the Rotate tool on the Editor toolbar.

3. Press the A key, type "180", and press Enter to rotate the selected building features 180 degrees.

The selected features are now oriented 180 degrees from their previous location.

Moving features

Now that the buildings are oriented properly, you are ready to move and scale them so that they fit inside the subdivision located near the bottom center of the map.

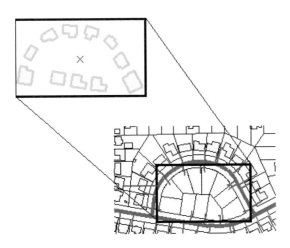

You can ensure the proper relocation of the building features by snapping the lower-left selected building feature to the endpoint of the lower-left water service line (shown in red).

1. With the buildings selected, click the Editor menu and click Snapping.

2. Check the End option for the Water layer and the Vertex option for the Buildings layer so that you can snap the corner of a building feature to the endpoint of a waterline.

3. Click the Edit tool so that you can move the selection anchor for selected features.

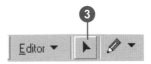

The selection anchor is a small x located at the center of selected features. It is the point on the feature or group of features that will be snapped when you move them.

4. Hold the Ctrl key down and move the pointer over the selection anchor. When the pointer icon changes, click and drag the selection anchor until it snaps to the corner of the lower-left building.

5. Drag the selected buildings until they snap to the endpoint of the waterline.

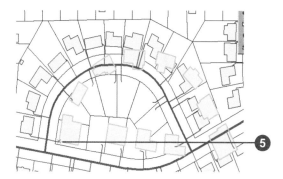

Notice that some of the buildings are too large to fit inside the parcels. You must scale these features to make them fit.

Scaling features

When data is created using a coordinate system different from that of your database, you may need to adjust the projection and scale of the data to fit the projection and scale of your database. Often, simply moving, rotating, and scaling those features are sufficient.

Because scaling is not a common operation, the Scale tool is not located on the Editor toolbar. You must therefore add it to the toolbar before you can use it.

1. Click the Tools menu and click Customize.

2. Click the Commands tab and click Editor in the Categories list. The Editor category contains all editing tools, regardless of their location.

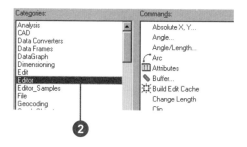

3. Scroll down the list of commands on the right until you find the Scale tool. Drag and drop the tool next to the Rotate tool on the Editor toolbar. Click Close on the Customize dialog box.

4. Before scaling the selected features, you may want to zoom in so that your scaling is more accurate. Click the Selection menu and click Zoom To Selected Features.

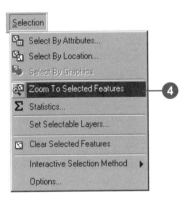

5. Click the Scale tool and drag the selected building features to scale them. Shrink the features until they fit inside the parcel subdivision. Use the waterlines as a guide. Scale features until the lower-right building matches the endpoint of the waterline.

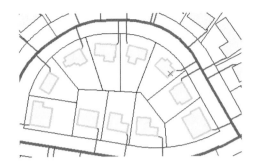

Extending and trimming waterlines using the Extend/Trim features task

Now that you have scaled the building features to fit inside the parcel subdivision, you need to extend the waterlines so that they snap to the side of each building. You can extend and trim waterlines using the Extend/Trim Features task.

1. To get a better view of the waterline that you need to extend, you can zoom in to the Extend Water Line bookmark. Click the View menu, click Bookmarks, and click Extend Water Lines.

2. Click the Current Task dropdown arrow and click Extend/Trim Features to set the edit task.

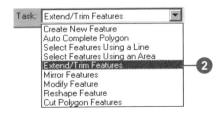

3. The Extend/Trim edit task will extend selected polyline features to the sketch you digitize. Click the Edit tool and click the waterline feature that you need to extend.

4. Click the Sketch tool and snap the first sketch point to the upper-right corner of the building feature you want to extend to.

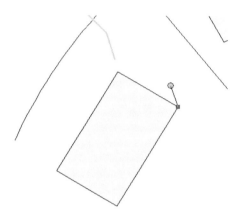

5. Move the pointer until it snaps to the upper-left building corner and double-click to finish the sketch. The waterline will then extend until it intersects the line that you have digitized. Since the line is identical to the side of the building, the end of the waterline should snap to the building.

You can also use the Extend/Trim features task to cut a waterline feature if it extends too far into the building.

6. To get a better view of the waterlines, you must zoom to the bookmarked extent called Trim Water Line, which was created for you. Click the View menu, point to Bookmarks, and click Trim Water Line.

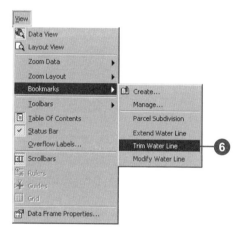

7. Click the Edit tool and click to select the waterline that extends into the building and needs to be trimmed.

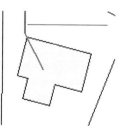

8. If you changed the current task, make sure that you change it back to Extend/Trim Features, then click the Sketch tool to start digitizing.

9. Snap the first sketch point to the lower-left corner of the building feature.

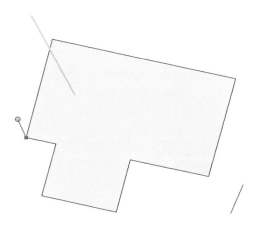

10. Move the pointer to the upper-left corner of the building. Double-click to snap the last point of the sketch to the building corner and trim the waterline feature.

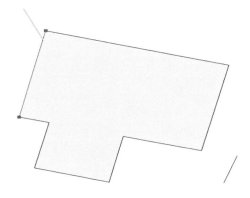

Extending and trimming waterlines using the Modify Features task

The Extend/Trim features task lets you extend and trim selected waterlines using a sketch that the features either cross or extend to. However, that is not the only method for extending or trimming waterlines. You can move, insert, or remove vertices of the waterline by making its shape the edit sketch. You can do this using the Modify Features task.

1. To get a better view of the waterlines, you need to zoom to the bookmarked extent called Modify Water Line. Click the View menu, point to Bookmarks, and click Modify Water Line.

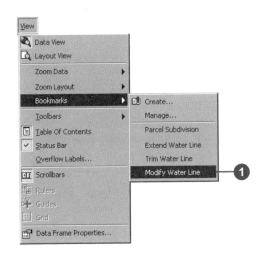

2. Click the Edit tool and click to select the waterline feature that needs to be extended.

3. Click the Current Task dropdown arrow and click Modify Feature to display the vertices of the waterline.

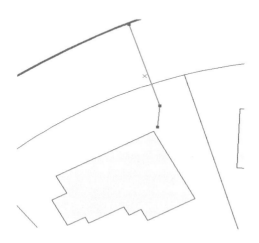

4. Click the Edit tool and move the pointer over the red vertex at the end of the waterline. Drag the vertex until it snaps to the building corner.

5. Move the pointer over the red vertex, right-click, then click Finish Sketch to finish modifying the waterline.

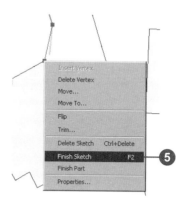

You can follow the same steps to trim line features using the Modify Features task. You can use the Trim command to reduce the length of the sketch by an exact distance as well.

With modifications to these waterlines completed, continue modifying the rest of the waterlines that don't connect to building features and experiment with other methods of modifying shapes.

For more information about editing features, see Chapter 7, 'Editing existing features'.

Exercise 5: Editing topological features

Most vector datasets have features that share common boundaries or points. Features that share common boundaries or points are said to have a topological association.

ArcMap lets you create and maintain topological associations between features and feature classes when you edit shapefiles, coverages, or feature datasets.

In the previous exercise, you learned how to create and manipulate individual features. In this exercise, you'll learn how to create and maintain topological associations between features.

Opening the exercise document

1. Start ArcMap.

2. Click the Open button on the Standard toolbar. Navigate to the EditingTopoFeatures.mxd map document located in the Editor directory where you installed the tutorial data (C:\ArcGIS\ArcTutor is the default location).

Integrating topological data

The dataset for this exercise contains state and county polygon layers for Ohio and West Virginia, as well as an Ohio River line layer.

The state, county, and river layers were taken from different data sources; they don't share common boundaries with each other. In order to update the boundaries that are shared between these features, you must create topological associations.

You'll begin by integrating all of the layers together so that all features with parts that should be shared are shared.

1. Click the Editor menu and click Start Editing.

The Integrate command lets you create topological associations between features. It makes all boundaries or vertices within a certain distance range identical or coincident. This distance range is called the *cluster tolerance*. You can choose to integrate only what you see or the entire dataset.

For this exercise, you must set the cluster tolerance to a value of 0.001 map units and integrate the entire dataset at that tolerance value.

2. Click the Editor menu and click Options. Click the Topology tab.

3. Uncheck the option to Integrate the visible extent only so that you can integrate the entire dataset.

4. Type "0.001" for the cluster tolerance value and click OK.

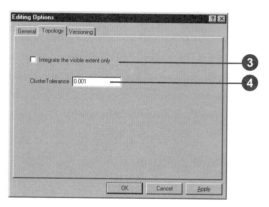

5. Click the Editor menu and click Integrate. This creates topological associations for the entire dataset.

Once you click Integrate, a progress indicator will appear. When the Integrate process is complete, you can analyze the result by zooming to the bookmarked extent called Integrate Result.

6. Click the View menu, point to Bookmarks, and click Integrate Result.

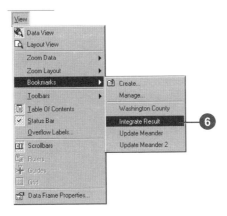

You can see your results by clicking the Undo button and then the Redo button at this zoomed extent. You should notice that the county and state boundaries are identical to the river that passes through them.

Reshaping a shared boundary

Now that the dataset has been integrated, you can start to modify the shared boundaries between features.

1. Click the View menu, point to Bookmarks, and click Update Meander.

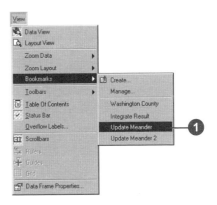

When the display updates, you should see a discrepancy between the boundary of the states and counties and the river layer. The meander, or bend, in the river layer is not accounted for in the state and county layers.

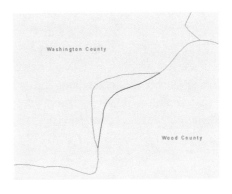

You can update the state and county layers so that they match the same boundary as the river feature by just reshaping the shared boundary between all of the state and county features. You can do this using the Shared Edit tool.

2. Click the Shared Edit tool.

3. Click the shared boundary you want to update.

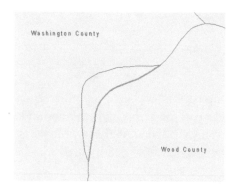

When you select a boundary with the Shared Edit tool, all the features in all the layers in the dataset that share this boundary are selected. This way, when you move part of a feature that is shared, any coincident and connected vertices underneath will move appropriately. This is true even for layers that aren't visible on the map.

With the shared boundary selected, you can update the state and county features to match the river line by using the Reshape Feature edit task.

4. Click the Current Task dropdown arrow and click Reshape Feature.

5. Click the tool palette dropdown arrow and click the Sketch tool.

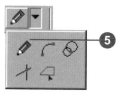

6. Right-click over the river line and click Replace Sketch.

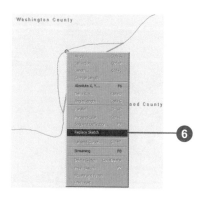

7. Right-click and click Finish Sketch or press F2.

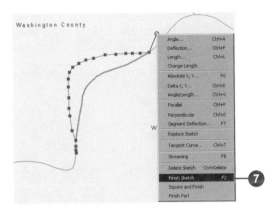

Both the Washington County and Wood County polygon features, as well as Ohio and West Virginia state polygon features, should now share the same meandering part of the river.

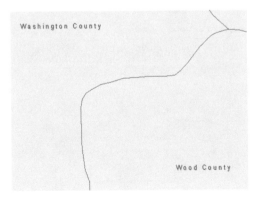

Modifying a shared boundary

Sometimes, updating a shared boundary is as simple as moving a couple of vertices. You can do this using the Modify Features task.

1. Click the View menu, point to Bookmarks, and click Update Meander 2.

2. Click the Shared Edit tool on the Editor toolbar, choose the shared state/county boundary, and double-click to update it.

Before you move any vertices of the shared boundary, set the snapping environment so that each vertex of the shared boundary snaps to the river line.

3. Click the Editor menu and click Snapping.

4. Check the Vertex option for the OhioRivers layer.

5. Click the Edit tool and drag the pointer over the second vertex until it snaps to the river line.

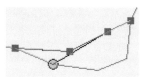

6. Right-click between the second and third vertex and click Insert Vertex.

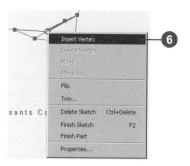

7. Click and drag the new vertex until it snaps to the bottom-left vertex of the river line.

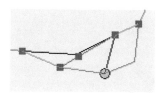

8. Move the second-to-last vertex until it snaps to the river line vertex located at the bottom left.

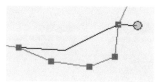

9. Right-click and click Finish Sketch or press F2 to finish editing the shared edge and update all features.

In this exercise, you used the topology tools in ArcMap to create topological associations between layers that represented the same geographic feature but had slightly different shapes. This is important when performing spatial analysis on layers that you attain from different sources.

With each task, you made sure that topological associations were maintained by only updating the shared boundaries between features.

For a more detailed discussion of the topics mentioned in this exercise, see Chapter 8, 'Editing topological features'.

Exercise 6: Working with CAD drawings

ArcMap lets you seamlessly integrate computer-aided design (CAD) drawings into your work. It allows you to display and query CAD datasets without first having to convert the drawing files to an ESRI format.

The ability to work with CAD drawings in ArcMap is particularly useful if your organization has existing CAD data resources that you need to use immediately in your work.

Not only can you perform basic query and analysis functions using ArcMap tools, but you can also snap directly to CAD features or entities when you update your database.

This exercise will show you how to import CAD features directly into your edit session; this will allow you to easily integrate CAD features into your work.

Opening the Exercise document

1. Start ArcMap.
2. Click the Open button on the Standard toolbar. Navigate to the WorkingWithCAD.mxd map document located in the Editor directory where you installed the tutorial data (C:\ArcGIS\ArcTutor is the default location).

3. Zoom to the area of the map identified by the red hashed polygon.

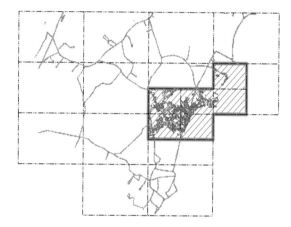

Using the Load Objects Wizard

You can import CAD entities directly from CAD feature classes using the Load Objects Wizard. However, you'll need to add the Load Objects Wizard into ArcMap first.

1. Click the Tools menu and click Customize. Click the Commands tab.

2. Click the Data Converters category from the Categories list and drag and drop the Load Objects command onto the Editor menu. Close the Customize dialog box.

3. Select the Editor menu and click Start Editing. Set the target layer to the LotLines layer. This is the layer into which you will load the parcel lines.

Loading CAD features

With the target layer set to the lot lines feature class, you are ready to load features directly from the CAD drawing.

CAD drawings are represented in two ways: CAD drawing files and CAD drawing datasets. CAD drawing datasets contain feature classes organized by point, line, or polygon shape types.

Each CAD feature in a CAD feature class contains a Layer field; it lets you identify the CAD drawing layer that each feature is derived from. In this exercise, you'll extract the features belonging to the lot line layer of the polyline feature class into your empty lot line geodatabase feature class.

1. Click the Editor menu and click Load Objects wizard.

2. Click the Browse button (located to the right of the Input data list). Navigate to where you installed the ArcTutor sample data (C:\Esri\ArcTutor by default), then navigate to the Editor\ExerciseData\EditingCAD directory.

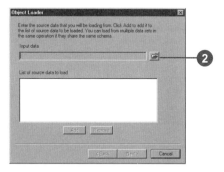

3. Double-click the Parcels.dwg drawing dataset. Choose the Polyline feature class and click the Open button.

4. Click the Add button to add the CAD feature class (listed in the Input data list) to the list of source data to load.

5. Click Next.

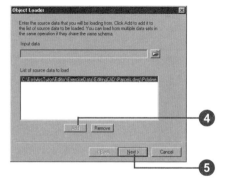

Matching input and target fields

The next step in the wizard lets you match the fields of the CAD feature class with the fields in your target layer.

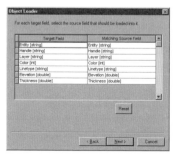

1. Accept the default field mappings for this exercise. Click Next.

Defining a query

Since all CAD layers are combined into a single feature class containing a Layer attribute value, you will define an attribute query so that only features with a layer name = 'LOT-L' will be loaded into the target layer.

1. Click the option to load only features that satisfy the query.

2. Click Query Builder to define the query.

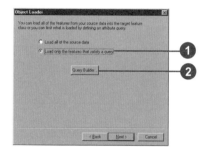

3. Double-click Layer in the Fields list. This adds the string to the where clause for the query.

4. Click the equals (=) sign.

5. Click Complete List to display all unique attribute values for the Layer field. Double-click LOT-L from the list to complete the query.

   After completing the steps above, your query should read: "Layer" = 'LOT-L'. You can alter the query by typing directly into the SQL query dialog box.

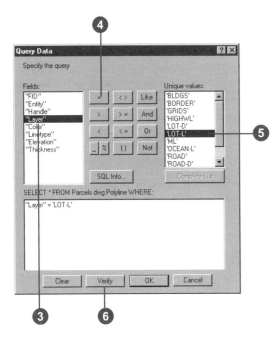

6. Click Verify to ensure that you have created a valid SQL where clause.

7. Click OK. Make sure that you have a valid query expression before applying the query to the wizard.

Snapping and validation

Next, the Object Loader will ask you if you want to apply any snapping agents that you have set in the Snapping Environment window to features as they are loaded into the map, and/or whether you want to validate each feature that is added.

If you're concerned about the connectivity between features that you import and existing features in your database, you may want to apply snapping. However, you should be aware that features will move within the current snapping tolerance. If the source CAD data was constructed using coordinate geometry, applying snapping may reduce the accuracy of the original data.

1. Click Next (do not apply snapping).

Completing the wizard and loading features

The final dialog box provides a summary of the options that you chose through each step of the wizard. You can examine each of your steps and click Back if you made any mistakes.

1. Click Finish.

   A progress indicator will appear.

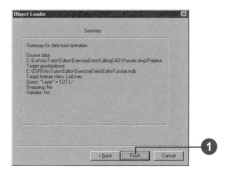

Once the wizard has finished loading features, you may need to refresh the display to see the new lot lines.

Saving your edits

Now that you have successfully loaded CAD data into your edit session, you can stop editing and save your edits.

1. Click the Editor menu and click Stop Editing.

2. Click Yes to save your edits.

In this exercise, you learned how to load CAD features directly into your GIS database. You were able to import features by their shape type and by their CAD layer name using the Load Objects wizard. But you don't have to import CAD data to use it. You can also snap directly to CAD features or simply display and query their attributes. For more information about CAD drawings, see *Using ArcCatalog*.

In the first two exercises, you learned how to use the edit sketch and sketch tools to create new features. There are a lot of additional methods for creating features that were not touched upon in these exercises. To learn about more ways to create new features, see Chapter 4, 'Creating new features'.

In addition to digitizing new features using the mouse, you learned how to use a digitizer puck and tablet to capture data from paper maps. Exercise 3 showed you how you can attach a paper map to your digitizing tablet, register the paper map to the coordinate space of your GIS database, and add features using the puck. To learn more about using a digitizing tablet, see Chapter 5, 'Using a digitizer'.

In exercise 4, you learned how easy it is to modify the shape of existing features. You copied and pasted buildings from a CAD file into your GIS database; you also moved, rotated, and scaled the buildings to match a parcel subdivision using some of the editing tools in ArcMap. Once the buildings were properly placed, you used the Extend/ Trim and Modify Feature edit tasks to connect water service lines to the side of each building. To learn more about editing features in ArcMap, see Chapter 7, 'Editing existing features'.

You can edit multiple features at the same time in ArcMap and ensure that the boundaries between them are consistent. In Exercise 5, you learned how to use the Integrate command to create common boundaries between state, county, and river features that were compiled from different sources at different levels of accuracy. You then updated both the state and county boundaries to match the changes (meanders) in the river features using the Shared Edit tool.

In the last exercise, you learned how to update your existing data with features in a CAD drawing file using the Load Objects wizard. You defined a query based on the lot line CAD layer type and then loaded only those features into your target layer.

Whether importing CAD data, using a digitizer to capture features from paper, or editing the shared boundaries between polygon features, ArcMap provides the tools you need to edit your data quickly and easily.

The next chapter will provide an overview of editing in ArcMap. It will show you where to find the tools that you need to use and describe the basic tasks you need to know before you start editing in ArcMap.

Editing basics

IN THIS CHAPTER

- An overview of the editing process

- Exploring the Editor toolbar

- Adding the Editor toolbar

- Adding the data you want to edit

- Starting and stopping an edit session

- Selecting features

- Moving features

- Deleting features

- Copying and pasting features

- Setting the number of decimal places used for reporting measurements

In addition to mapmaking and map analysis, ArcMap is also the application for creating and editing your spatial databases. ArcMap has tools to edit shapefiles, coverages, and feature datasets—any type of vector data in your GIS database.

This chapter provides an introduction on how to edit in ArcMap and describes the basic tasks you need to know before you can start to create and edit spatial data. For instance, this chapter shows you how to perform such tasks as adding the Editor toolbar, starting and stopping an edit session, and selecting features.

An overview of the editing process

The following is a general overview of how to use ArcMap and the Editor toolbar to edit your data. Each of the following steps is outlined in detail in this chapter or other chapters in this section.

1. Start ArcMap.

2. Create a new map or open an existing one.

Open button

New map file button

3. Add the data you want to edit to your map.

Add data button

If there are no existing layers for the feature classes you want to edit, you can create them using ArcCatalog™. For more information on creating a feature layer, see *Using ArcCatalog*.

4. Add the Editor toolbar to ArcMap.

Editor Toolbar button

5. Choose Start Editing from the Editor menu.

6. Create or modify features and/or their attributes.

7. Choose Stop Editing from the Editor menu and click Yes when prompted to save your edits.

There is no need to save the map—all edits made to the database will automatically be reflected the next time you open the map.

The Editor toolbar

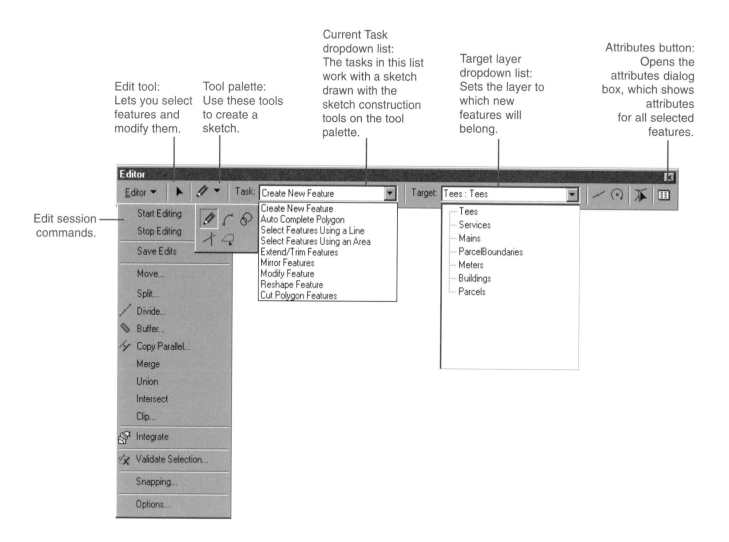

Edit tool:
Lets you select features and modify them.

Tool palette:
Use these tools to create a sketch.

Current Task dropdown list:
The tasks in this list work with a sketch drawn with the sketch construction tools on the tool palette.

Target layer dropdown list:
Sets the layer to which new features will belong.

Attributes button:
Opens the attributes dialog box, which shows attributes for all selected features.

Edit session commands.

Editor

Editor ▾   Task: Create New Feature   Target: Tees : Tees

Start Editing
Stop Editing
Save Edits

Move...
Split...
Divide...
Buffer...
Copy Parallel...
Merge
Union
Intersect
Clip...
Integrate
Validate Selection...
Snapping...
Options...

Create New Feature
Auto Complete Polygon
Select Features Using a Line
Select Features Using an Area
Extend/Trim Features
Mirror Features
Modify Feature
Reshape Feature
Cut Polygon Features

Tees
Services
Mains
ParcelBoundaries
Meters
Buildings
Parcels

Exploring the Editor toolbar

This section shows you how editing in ArcMap helps you complete the tasks that you need to do. You'll learn about the types of data you can edit as well as the basics of creating and modifying features and their attributes.

The structure of vector datasets

ArcMap provides a common editing environment for features stored in all types of geographic datasets: feature datasets, coverages, and shapefiles.

When you edit data with ArcMap, you edit feature classes (collections of features) that the layers on your map represent.

Editing the feature classes lets you edit the actual data source, not just the representation on the map.

A feature class is a collection of the same type of features, for example, a collection of points or a collection of polygons. For each type of geographic dataset, the available types of features vary (see table below).

A dataset is a collection of feature classes that share the same spatial reference. A dataset might be a collection of land base feature classes or a collection of utility feature classes. Shapefiles are an exception; they do not hold a collection of feature classes, but only one shapefile feature class.

Comparing the structure of vector datasets

	Geodatabase	Coverage	Shapefile
Collections of datasets	A *geodatabase* is a collection of feature datasets.	An *ArcInfo workspace* is a collection of coverages.	A shapefile folder is a collection of shapefiles.
Datasets	A *feature dataset* is a collection of feature classes.	A *coverage* is a collection of coverage feature classes.	A *shapefile* has one shapefile feature class.
Collections of features	A *feature class* is a collection of features of the same type.	A coverage feature class is a collection of coverage features.	A shapefile feature class is a collection of shapefile features.
Features	Point, multipoint, polyline, polygon, annotation, and network.	Primary coverage feature classes: point or label point, arc, and node. Secondary feature classes: polygon, tic, link, section, and annotation. Compound feature classes: region and route.	Point, multipoint, line, and polygon.

A collection of feature datasets is stored in a geodatabase. Coverages are stored in an ArcInfo workspace, and shapefiles are stored in a shapefile folder. Although you may add multiple collections of datasets to your map (geodatabases, ArcInfo workspaces, and shapefile folders), you can only edit feature classes within one collection at a time. Also, some coverage feature classes can't be edited—link, section, annotation, and group feature classes.

What is a sketch and how does it work with a task?

A *sketch* is a shape you draw that performs various tasks when editing such as adding new features, modifying features, and reshaping features. Tasks are listed in the Current Task dropdown list. You must create a sketch in order to complete a task.

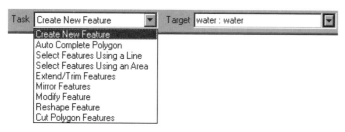

Current Task dropdown list

For instance, the Create New Feature task uses a sketch you create to make the new feature.

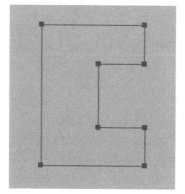

Building as sketch

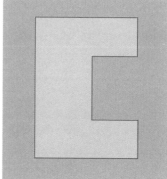

Building as feature

The Select Features Using a Line task uses a sketch you create to select features; the features the line intersects are selected.

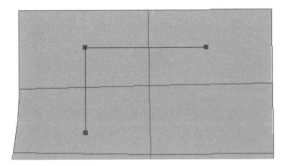

Sketch intersects parcels to be selected.

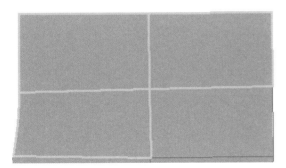

Parcels intersected by the sketch are now selected.

The Cut Polygon Features task uses a line sketch you draw to cut a polygon.

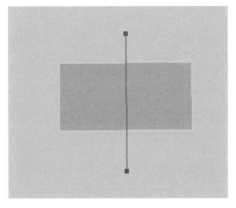

Sketch showing where the polygon is to be "cut".

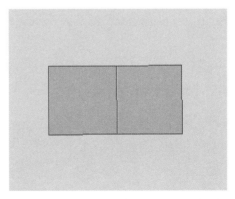

Polygon divided into two features where the sketch was drawn.

Creating new features

You can create three main types of features with the Editor toolbar: points, lines, and polygons.

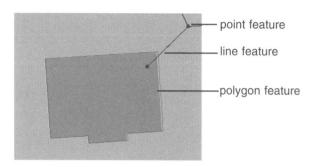

point feature

line feature

polygon feature

To create a line or polygon, you must first create a sketch. A sketch's shape is composed of all the vertices and segments of the feature. *Vertices* are the points at which the sketch changes direction, such as corners, and *segments* are the lines that connect the vertices.

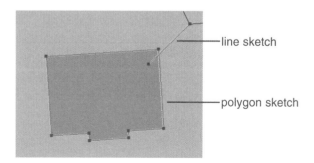

line sketch

polygon sketch

You can create a sketch by creating the vertices and segments that make up the features. Vertices are marked in green, with the last vertex added marked in red.

The Sketch tool is the tool you use most often to create a sketch. It has an accompanying context menu that helps you place vertices and segments more accurately. The Arc tool, the Distance–Distance tool, and the Intersection tool (located with the Sketch tool on the tool palette) also help you create vertices and segments using other construction methods.

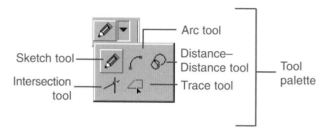

Arc tool

Sketch tool

Distance–Distance tool

Intersection tool

Trace tool

Tool palette

Angle...	Ctrl+A
Deflection...	Ctrl+F
Length...	Ctrl+L
Change Length	
Absolute X, Y...	F6
Delta X, Y...	Ctrl+D
Angle/Length...	Ctrl+G
Parallel	Ctrl+P
Perpendicular	Ctrl+E
Segment Deflection...	F7
Replace Sketch	
Tangent Curve...	Ctrl+T
Streaming	F8
Delete Sketch	Ctrl+Delete
Finish Sketch	F2
Square and Finish	
Finish Part	

Sketch tool context menu

When you're creating a new feature, the target layer determines in which layer a new feature will belong. The Target layer dropdown list contains the names of all the layers in the datasets with which you're working. Subtypes are also listed, if applicable. For instance, if you set the target layer to "Buildings: Commercial", any features you create will be part of the "Commercial" subtype of the "Buildings" layer.

Target layer dropdown list

You must set the target layer whenever you're creating new features—whether you're creating them with the Sketch tool, by copying and pasting, or by buffering another feature.

Modifying features

For every feature on the map, there is an alternate form, a sketch. In the same way that you must create a sketch to create a feature, to modify a feature you must modify its sketch. Because the vertices are visible in a sketch, you can edit the feature in detail; you can move the vertices, delete them, or add new ones using the Sketch context menu.

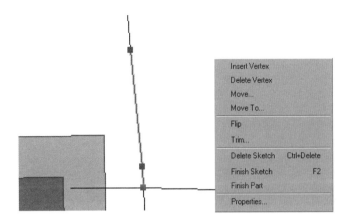

When you edit a feature's sketch, you edit its vertices using the Sketch context menu.

Besides editing a feature by working with its sketch, you can also use another sketch you create to modify the feature for certain tasks. An example of this type of task is Cut Polygon Features, where a sketch you construct is used to divide one polygon into two.

Simple modifications to features, such as moving, copying, or deleting, can be made by selecting the feature and choosing the appropriate tool or command.

Editing attributes

Attributes can be created or edited in the Attributes dialog box. After selecting the features whose attributes you want to edit, click the Attributes button to see the dialog box.

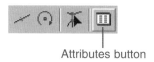

Attributes button

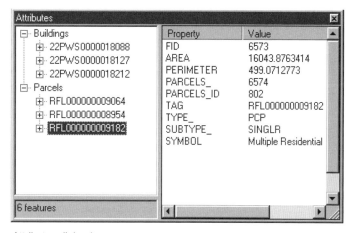

Attributes dialog box

Adding the Editor toolbar

Before editing geographic feature data within ArcMap, you must first add the Editor toolbar.

Tip

Adding the Editor toolbar from the Tools menu
You can also add the Editor toolbar from the Tools menu. Click Tools and click Editor Toolbar.

Tip

Adding the Editor toolbar from the View menu
You can also add the Editor toolbar by clicking the View menu, pointing to Toolbars, and checking Editor.

Tip

Adding the Editor toolbar using the Customize dialog box
Click the Tools menu and click Customize. Click the Toolbars tab and check Editor.

1. Start ArcMap.

2. Click the Editor Toolbar button on the ArcMap Standard toolbar to display the Editor toolbar.

3. Click the toolbar's title bar and drag it to the top of the ArcMap application window.

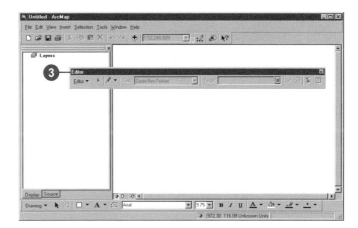

Adding the data you want to edit

Before you can start editing, you must add the data you want to edit to your map. In ArcMap, you can edit datasets in shapefile, coverage, or geodatabase format.

Tip

Stopping the drawing of data

You can stop the drawing process without clearing the map by pressing the Esc key.

Tip

Loading data from a geodatabase

You can import features from a geodatabase into a layer on your map using the Load Objects command. For more information, see Building a Geodatabase.

1. Start ArcMap.

2. Click the Add Data button.

3. Navigate to the location of your data and click Add.

   The data is added to your map.

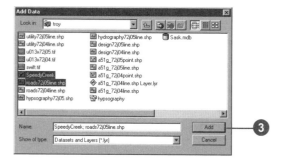

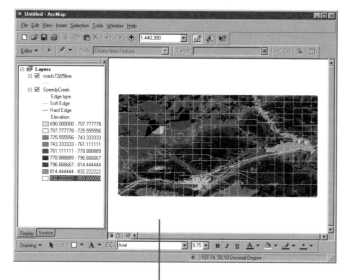

Data is added to the map.

Starting and stopping an edit session

All editing takes place within an *edit session.* To begin, choose Start Editing from the Editor menu. The edits you make are immediately visible on your map but are not saved to the database until you choose to do so.

If you're working with large amounts of data, you can speed up the editing and selection of features by creating an edit cache. An *edit cache* holds the features visible in the current map extent in memory on your local machine. An edit cache results in faster editing because ArcMap doesn't have to retrieve data from the server. You can create an edit cache by clicking the Build Edit Cache command on the Edit Cache toolbar. ►

Tip

Editing a map with more than one collection of datasets

You can only edit one collection of datasets—one workspace—at a time. These can be geodatabases, ArcInfo coverages, and shapefiles. If your map contains more than one collection, when you choose Start Editing you will be prompted to choose which one you want to edit.

Starting an edit session

1. Start ArcMap and add the Editor toolbar.

2. Click Editor and click Start Editing.

   The Editor toolbar is now active.

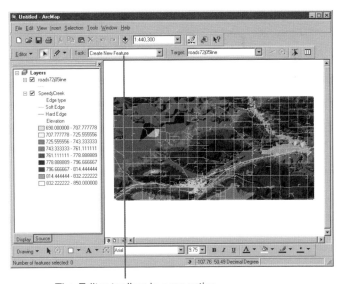

The Editor toolbar is now active.

When you're finished editing, you can save any changes you've made or quit editing without saving. You can also save the edits you've made at any time by choosing Save Edits from the Editor menu.

Zooming to your edit cache extent

You can quickly return to your edit cache extent at any time in your edit session. Click the Zoom to Edit Cache button on the Edit Cache toolbar.

Creating an edit cache

1. Add the data you want to edit.

2. Click the Zoom In button on the Tools toolbar.

3. Zoom in to the area on the map that you want to edit.

4. Click the Build Edit Cache cache button on the Edit Cache toolbar.

   The features visible in the current extent are held in memory locally.

Saving your edits in the middle of an edit session

1. Click Editor.

2. Click Save Edits.

   Any edits you have made are saved to the database.

Editing a map with more than one data frame

If your map contains more than one data frame, you will be editing the data frame that is active when you choose Start Editing. To edit a different data frame, you must choose Stop Editing and then choose Start Editing with the desired data frame active.

For a discussion of data frames, see 'Layers, data frames, and the table of contents' in Chapter 3 of Using ArcMap.

Editing in layout view

You can also edit data in a map that you're preparing. Click the View menu and click Layout View. For more information about working in layout view, see 'Editing in data view and layout view' in Chapter 1.

Stopping an edit session

1. Click Editor and click Stop Editing.

2. To save changes, click Yes. To quit without saving, click No.

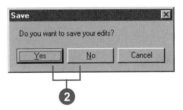

Selecting features

Selecting features identifies the features on which you want to perform certain operations. For example, before you move, delete, or copy a feature, you must select it. You must also select features before you can view their attributes.

You can select features in several different ways, either by clicking them with the Edit tool or by creating a line or a polygon that intersects the features you want to select. The number of features selected is shown immediately after you make the selection, in the lower-left corner of the ArcMap window. ▶

Tip

Selecting more than one feature

To select more than one feature, hold down the Shift key while you click the features. You can also use the Edit tool to drag a box around a group of features.

Tip

Removing features from the selection

To remove features from the selection set, hold down the Shift key while you click the features.

Selecting features using the Edit tool

1. Click the Edit tool.
2. Move the pointer over a feature and click the mouse.

   The selected feature is highlighted.

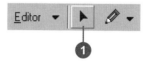

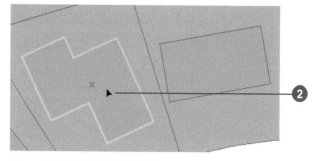

The selected building is highlighted.

The small "x" located in the center of the selected features is the *selection anchor*. The selection anchor is used when rotating features, moving features using snapping, and scaling features.

You can choose which layers you can select by choosing Set Selectable Layers from the Selection menu and using the Selectable Layers list.

For example, suppose you wanted to select a large number of buildings by drawing a box around them but selected a parcel by mistake as you drew the ►

Tip

Moving the selection anchor

To move the selection anchor, move the pointer over it, press the Ctrl key, and drag the selection anchor to the desired location.

See Also

For more information on the selection anchor, see 'Moving features' in this chapter and 'Scaling features' in Chapter 7.

See Also

For more information on creating a line, see 'Creating lines and polygons' in Chapter 4.

Selecting features using a line

1. Click the Current Task dropdown arrow and click Select Features Using a Line.

2. Click the tool palette dropdown arrow and click the Sketch tool or any of the other construction tools in the tool palette.

3. Construct a line that intersects the features you want to select.

   The features that the line intersects are now selected.

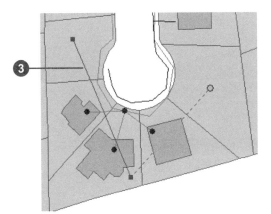

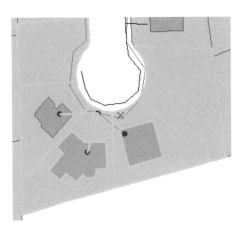

Features intersected by the line are now selected.

selection box. To avoid this, you might uncheck the Parcels layer in the Selectable Layers list so that parcels cannot be selected.

Tip

Use the Selection menu for more control over your selection

The Selection menu gives you more tools with which to make a selection, such as adding to the current selection, selecting all features onscreen, or creating an SQL statement.

Tip

Pan and zoom while you select features

You can pan and zoom while selecting features without having to change tools. Hold the Control key and press Z to zoom in, X to zoom out, or C to pan.

See Also

For more information on creating a polygon, see 'Creating lines and polygons' in Chapter 4.

See Also

For more information on selecting features in ArcMap, including selecting features for analysis, see Chapter 13, 'Querying maps', in Using ArcMap.

Selecting features using a polygon

1. Click the Current Task dropdown arrow and click Select Features Using an Area.

2. Click the tool palette dropdown arrow and click the Sketch tool or any of the other construction tools in the tool palette.

3. Construct a polygon that intersects the features you want to select.

   The features that intersect with the polygon you created are now selected.

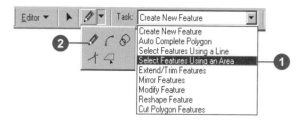

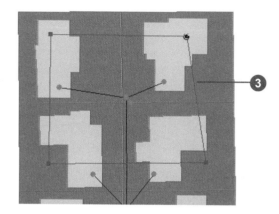

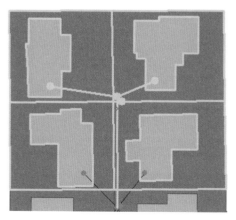

Features that intersect with the area are now selected.

Making a layer selectable

1. Click Selection and click Set Selectable Layers.

2. Click the check boxes next to the layer names you want to be able to select. Uncheck the boxes next to the names you don't want to be able to select.

   Layers whose names are unchecked are still visible in your map but cannot be selected.

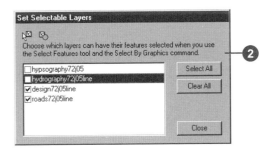

Moving features

You can move features in three different ways: by dragging; by specifying delta x,y coordinates; or by rotating.

Dragging is the easiest way to move a feature. Use this method when you have a general idea of where you want to move the feature.

Specify delta x,y coordinates when you want to move a feature to a precise location. ArcMap uses the current location of the selected feature or features as the origin (0,0) and moves them from that location according to the coordinates you specify. ▶

Tip

"Undoing" a move

You can undo any edit you make to a feature by clicking the Undo button on the ArcMap Standard toolbar.

Dragging a feature

1. Click the Edit tool.
2. Click the feature or features you want to move.
3. Click and drag the feature or features to the desired location.

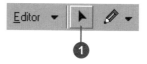

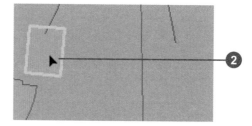

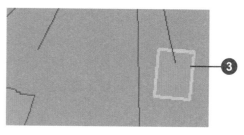

The selected building is dragged to a new location.

The coordinates are measured in map units. The graphic below illustrates the change in location when delta x,y coordinates of 2,3 are specified for a building.

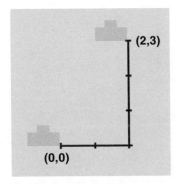

In the example above, the building is moved up and to the right, as positive coordinate values are specified. To move the building left and down, you would specify negative values.

You can rotate features in ArcMap using the Rotate tool. After selecting the features, drag the mouse pointer so that the features rotate to the desired position. Features rotate around the selection anchor, the small "x" located in the center of selected features.

If you want to move a feature to a precise location in relation to another feature, you can use the snapping environment. For example, you can move a ►

Moving a feature relative to its current location

1. Click the Edit tool.

2. Click the feature or features you want to move.

3. Click Editor and click Move. ►

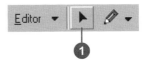

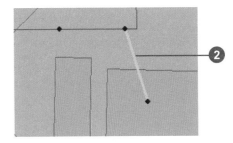

parcel and have one of its corners jump, or "snap", precisely to a corner of another parcel. Simply move the parcel's selection anchor to its corner vertex after setting the appropriate snapping properties. Then, move the parcel toward its new location until the selection anchor snaps to the corner vertex of the other parcel. Snapping is discussed in detail in Chapter 4, 'Creating new features'.

Tip

Moving the selection anchor

To move the selection anchor, move the pointer over it, press the Ctrl key, and drag the selection anchor to the desired location.

4. Type the desired coordinates and press Enter.

   The feature is moved according to the specified coordinates.

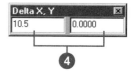

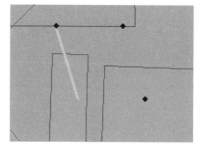

The feature is moved 40 map units to the left.

Rotating by degree

To specify the exact amount of counterclockwise rotation, click the Rotate tool, press A, and type the number of degrees. A positive number rotates the feature to the right, a negative number to the left.

Rotating a point's symbology

If your data already has a field that contains the rotation angle for each point symbol, you can use ArcMap to rotate the symbology.

Right-click the point layer name in the map's table of contents and click Properties. Click the Symbology tab. Click the Advanced button, then click Rotation. From the dropdown list, choose the field that contains the rotation angle. Click the option that describes how you want that angle calculated.

Rotating a feature

1. Click the Edit tool.
2. Click the feature or features you want to rotate.
3. Click the Rotate tool.
4. Click anywhere on the map and drag the pointer to rotate the feature to the desired position.

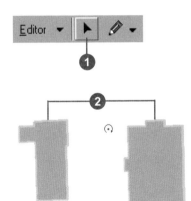

The selected features are rotated.

Deleting features

To delete a feature from the map and from the database, use the Delete button on the ArcMap Standard toolbar.

Tip

Deleting features using the Delete key

You can also press the Delete key on the keyboard to remove selected features.

1. Click the Edit tool.
2. Click the feature or features you want to delete.
3. Click the Delete button on the ArcMap Standard toolbar.

   The selected features are deleted.

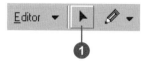

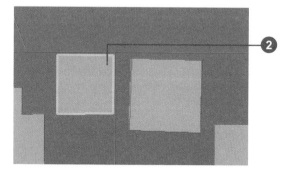

The selected building is deleted.

Copying and pasting features

To copy an existing feature, use the tools on the ArcMap Standard toolbar. From the Target layer dropdown list, choose the layer containing the type of features you want the new feature to be—for example, a building.

You can copy a feature and paste it as part of another layer, but it must be the same type of layer (point, line, or polygon) as the one from which you copied. There is one exception to this rule—you can copy polygons into a line layer.

Attributes from the original feature are only copied to the new feature if you are copying and pasting within the same layer.

See Also

For more information on attributes, see Chapter 9, 'Editing attributes'. You can also see Chapter 10 of Using ArcMap.

1. Click the Target layer dropdown arrow and click the layer containing the type of features you want the new features to be.

2. Click the Edit tool.

3. Click the feature or features you want to copy.

4. Click the Copy button on the ArcMap Standard toolbar.

5. Click the Paste button on the ArcMap Standard toolbar.

   The feature is pasted on top of the original feature.

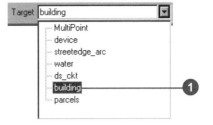

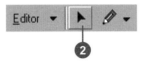

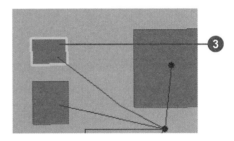

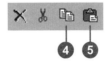

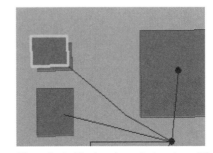

The selected feature is pasted on top of the original feature.

Setting the number of decimal places used for reporting measurements

When creating or editing a feature with the Sketch tool, you can use the Sketch tool context menu to view such measurements as the distance between two vertices, the angle between two segments, or the current coordinate location of the pointer.

By default, ArcMap displays these measurements using four decimal places. However, you can easily change the number of decimal places displayed. After you set the number of decimal places, ArcMap will report all measurements using that number of decimal places.

1. Click Editor and click Options.

2. Click the General tab.

3. Type the number of decimal places you want to use.

4. Click OK.

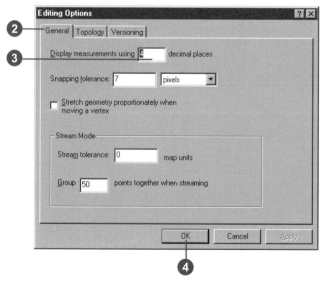

Creating new features

4

It's easy to create new features using the editing sketch construction tools. First, specify the layer in which you want to create the new feature. Then, use the appropriate tool to digitize the vertices of the feature.

You can use the editing tools to create new point, line, or polygon features for many practical purposes. Use the Sketch tool and its accompanying context menu to add a water main perpendicular to an existing water main in a subdivision. Use the Distance–Distance tool to create a land parcel that begins 55 meters from one corner of an existing lot and 40 meters from another lot corner. Create a cul-de-sac using the Arc tool to create a parametric (true) curve. With the Intersection tool, add a parcel to a subdivision by establishing a corner vertex using segments of an adjoining parcel.

Using the editing tools, you can create a variety of features by constructing segments at specific angles and of specific lengths. You can create features that are parallel or perpendicular to other features. You can also create multipoint features, such as a system of oil wells, and multipart features, such as a group of islands that form a country or state.

These are just a few examples of how you can use ArcMap to easily and accurately create new features for your database.

How to create a new feature

To create a new feature using ArcMap, you create an edit sketch. A *sketch* is a shape that you draw by digitizing vertices. You can use a sketch to complete various tasks; these tasks are listed in the Current Task dropdown list shown below. Tasks you can complete with a sketch include creating new features, modifying features, extending or trimming features, and reshaping features.

Current Task dropdown list

This chapter focuses on using sketches to create new features. When the current task setting is Create New Feature, the shape you create becomes the new feature.

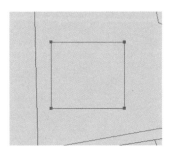

Building as sketch

Building as feature

A sketch is composed of *vertices* (the points at which the sketch changes direction such as corners) and *segments* (the lines that connect the vertices). You create a sketch using the Sketch tool located on the tool palette.

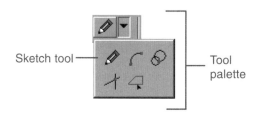

Sketch tool — Tool palette

To create point features, click once on the map. To create line or polygon features (see the example below), use the Sketch tool to click on the map to digitize the vertices that make up that feature. To create the last vertex and finish the sketch, double-click with the mouse. After you finish the sketch, ArcMap adds the final segment of the sketch and the sketch turns into a feature.

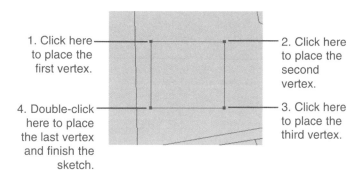

1. Click here to place the first vertex.

2. Click here to place the second vertex.

3. Click here to place the third vertex.

4. Double-click here to place the last vertex and finish the sketch.

The type of feature you create is determined by the setting of the Target layer dropdown list. This list contains the names of all the layers in the datasets with which you're working.

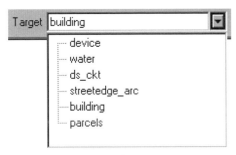

Target layer dropdown list

Choose the layer to which you want to add new features before you start to create them.

Of course, you won't always be able to place vertices or segments interactively. When you're using the Sketch tool, you can see a menu called the Sketch tool context menu. You can access this menu when you right-click the mouse away from the sketch you're creating. The menu has choices to help you place the vertices and segments exactly where you want them. For example, you can set a segment to be a certain length or angle or create a vertex at a specific x,y coordinate location.

Sketch tool context menu

All the tools on the tool palette help you create a sketch. Two tools use more specific construction methods to create either points or vertices: the Distance–Distance tool and the Intersection tool.

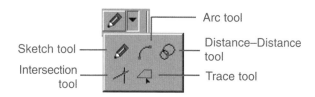

Arc tool

Sketch tool

Distance–Distance tool

Intersection tool

Trace tool

The Distance–Distance tool lets you create a point or vertex at the intersection of two distances from two other points. You might use this tool to place a new electrical primary based on field measurements. Suppose you know that the next point for the primary is 50 feet from one building corner and 75 feet from another.

The Distance–Distance tool creates two circles based on these distances and finds two possible intersection points where the primary can be placed.

The Intersection tool creates a point or vertex at the place where two segments would intersect if extended far enough. Suppose you want to create a parking lot adjoining an L-shaped building. The outer corner of the lot should be located at the point where the two outermost walls of the building would intersect if they were extended. You could use the Intersection tool to find this "implied" intersection point and create the corner vertex of the lot (see graphic below).

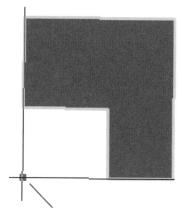

The Intersection tool creates a vertex here—
at the place where the two segments would intersect.

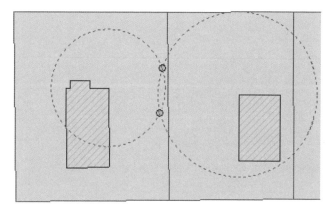

The Distance–Distance tool gives the intersection points of two circles; the size of the circles is determined by the radius you set.

The Arc tool helps you create a segment that is a parametric (true) curve.

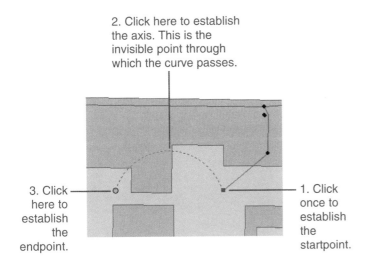

2. Click here to establish the axis. This is the invisible point through which the curve passes.

3. Click here to establish the endpoint.

1. Click once to establish the startpoint.

The Trace tool helps you create segments that follow along existing segments. Suppose you want to add a new road casing feature that is offset 15 feet from the front of a parcel subdivision. You could use the Trace tool to trace along the existing line features instead of typing the angle and length of each segment.

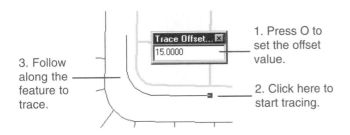

3. Follow along the feature to trace.

1. Press O to set the offset value.

2. Click here to start tracing.

You can use any combination of the following methods for creating vertices or segments to create a new line or polygon feature:

- Sketch tool
- Sketch tool context menu
- Distance–Distance tool
- Intersection tool
- Arc tool
- Trace tool

ArcMap has another context menu—the Sketch context menu—that works more directly with the sketch as a whole. With this menu, you can add, move, or delete vertices; switch the direction of the sketch; reduce its length; or display the properties of the sketch shape. From the properties dialog box, you can remove parts from a multipart feature, remove many vertices in one operation, add points, and/or modify m- and z-values. The Sketch context menu is available when you right-click while the pointer is positioned over any part of the sketch using any tool. It differs from the Sketch tool context menu, which you can access only when working with the Sketch tool and when you right-click away from your sketch.

Sketch context menu

Creating point features and vertices

You can think of vertices as being much the same as point features, except that vertices are connected by segments and make up line or polygon features.

Point features and vertices are created using the same methods. The Target layer setting determines whether you're creating a point feature or a vertex that is part of a line or polygon sketch.

You can create point features or vertices of a sketch in several different ways:

- By digitizing freehand with the Sketch tool (you can also use the snapping environment to help)
- By using Absolute X, Y or Delta X, Y on the Sketch tool context menu ▶

Tip

The snapping environment can help you create points and vertices

The snapping environment can help you create points or vertices at more exact locations relative to other features. For more information, see 'Using the snapping environment' in this chapter.

Creating a point or vertex by digitizing

1. Click the Current Task dropdown arrow and click Create New Feature.

2. Click the Target Layer dropdown list and click a point layer.

3. Click the tool palette dropdown arrow and click the Sketch tool.

4. Click on the map to create the point.

   The point or vertex is created on your map and marked as selected.

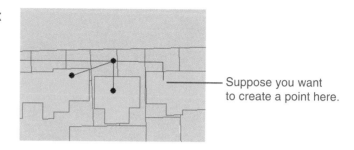

Suppose you want to create a point here.

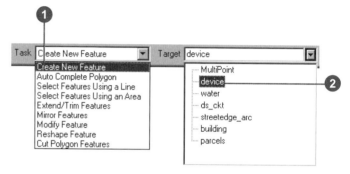

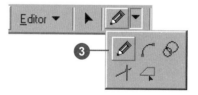

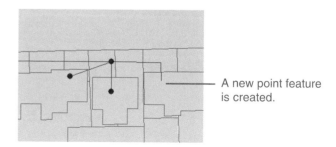

A new point feature is created.

- By using the Distance–Distance or Intersection tools

To digitize freehand, simply click the Sketch tool and click on the map.

Absolute X, Y on the Sketch tool context menu lets you create a point or vertex using the origin (0,0) of the map's coordinate system. In other words, the location of the point or vertex you're creating is determined using the same point for 0,0 as your map data. You might use Absolute X, Y to create a pole in a utility database if you have the x,y coordinates of a pole from using a global positioning system (GPS) unit. ▶

Tip

Shortcut for Absolute X, Y
After clicking the Sketch tool, you can press F6 to set the x,y coordinates.

Tip

Closing the Sketch tool context menu
You can close the Sketch tool context menu by pressing the Esc key.

Creating a point or vertex using the coordinate system of the map (Absolute X, Y)

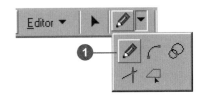

1. Click the tool palette dropdown arrow and click the Sketch tool.

2. Right-click anywhere on the map and click Absolute X, Y.

3. Type the coordinates and press Enter.

   A vertex or point is created at the specified coordinates.

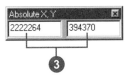

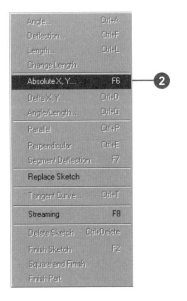

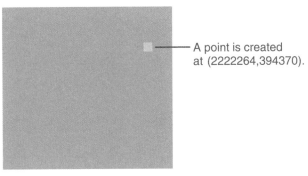

A point is created at (2222264,394370).

Delta X, Y on the Sketch tool context menu lets you create a vertex using the last vertex in the sketch as the origin. You can think of it as another way of measuring angle and length from a point already on the map.

For example, just as the red point in the diagram below can be measured at a distance of 20 feet from the last point at an angle of 45 degrees, it can also be measured in coordinates measured from the last point. ▶

20 ft.

45°

Point measured using an angle and length

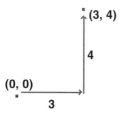

(3, 4)

4

(0, 0)

3

Same point measured using delta x,y coordinates

Tip

Shortcut for Delta X, Y
After clicking the Sketch tool, you can press Ctrl + D to set the delta x,y coordinates.

Creating a vertex relative to the location of the last vertex (Delta X,Y)

1. Click the tool palette dropdown arrow and click the Sketch tool after creating at least one vertex.

2. Right-click away from the vertex or sketch and click Delta X, Y.

3. Type the coordinates and press Enter.

   A vertex is created at the specified coordinates.

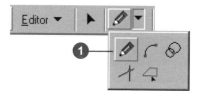

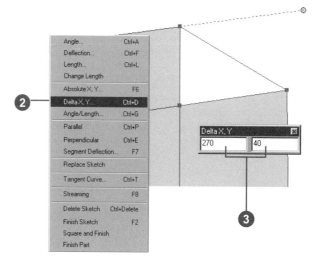

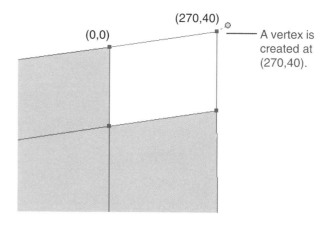

A vertex is created at (270,40).

The Distance–Distance tool offers another way to create a point or vertex at a specific location. Suppose you want to create a pole feature. If you don't have the exact coordinate location but know that it is at the intersection of 50 map units from the corner of one building and 70 map units from the corner of another, you can use the Distance–Distance tool to place the point. The Distance–Distance tool lets you create a point or vertex at the intersection of two distances from two other points.

As shown in the example, you'd create one circle with the centerpoint on the corner of the first building and a radius of 50 map units. You'd create another circle with the centerpoint on the corner of the other building and a radius of 70 map units. The Distance–Distance tool calculates the two locations where the radii of the circles intersect.

The Intersection tool creates a point or vertex at the implied intersection of two segments. "Implied" means that the segments don't have to actually intersect on the map. The Intersection tool creates a point or vertex at the place where the segments would intersect if extended far enough.▶

Creating a point or vertex using the Distance–Distance tool

1. Click the tool palette dropdown arrow and click the Distance–Distance tool.

2. Click once to establish the centerpoint of the first circle and press the letter D on the keyboard.

3. Type the radius length for the first circle and press Enter.

   A circle is created with the specified radius. ▶

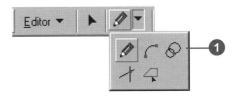

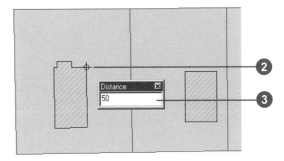

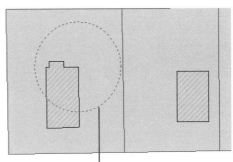

A circle with a 50-map unit radius is created.

For example, suppose you want to create a new parcel. One corner of the parcel must be placed at the implied intersection of two segments of an adjoining parcel. You can use the Intersection tool to find this implied intersection point and create the corner vertex of the new parcel.

You can also create a *multipoint feature*, a feature that consists of more than one ▶

Tip

Choosing an intersection point

Press Tab to alternate between the two points of intersection and press Enter to create the point.

Tip

"Undoing" and "redoing" a vertex

You can undo any vertex you create by clicking the Undo button on the ArcMap Standard toolbar. Click the Redo button if you want to readd the vertex.

4. Click once to establish the centerpoint of the second circle and press the letter D on the keyboard.

5. Type the radius length for the second circle and press Enter.

   A second circle is created with the specified radius. The two locations where the radii of the circles intersect are highlighted when you move the pointer over them.

6. Position the pointer over the location you want and click.

   A vertex or point is added to your map.

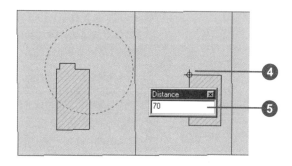

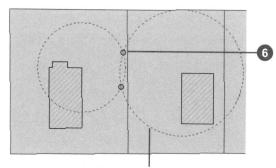

A circle with a 70-map unit radius is created.

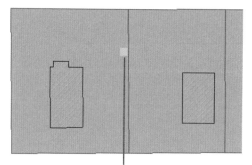

A point is created at one of two places where the radii of the circles intersect.

point but only references one set of attributes in the database. A system of oil wells is an example of a multipoint feature; the database references a single set of attributes for the main well and the multiple well holes in the system.

To create new features, you must have an existing layer to which you want to add them. If you do not, you can create one using ArcCatalog™. For more information on creating a feature layer, see *Using ArcCatalog*.

Tip

Pan and zoom while adding points

Hold down the Control key and press Z to zoom out, X to zoom in, or C to pan the display when using any sketch tool.

Creating a point or vertex using the Intersection tool

1. Click the tool palette dropdown arrow and click the Intersection tool.

   The pointer turns into crosshairs.

2. Position the crosshairs over the first segment and click.

3. Position the crosshairs over the second segment and click.

   A vertex or point is added at the implied intersection of the two segments.

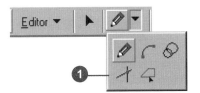

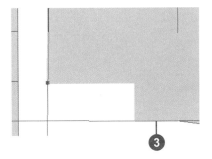

A vertex is added at the place where the two segments would intersect if extended.

Creating a multipoint feature

1. Click the Current Task dropdown arrow and click Create New Feature.

2. Click the Target layer dropdown arrow and click a multipoint layer.

3. Click the tool palette dropdown arrow and click the Sketch tool.

4. Click on the map to create parts of the multipoint feature.

5. When you have created the last point of the multipoint feature, right-click anywhere on the map and click Finish Sketch. ▶

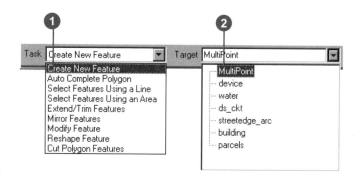

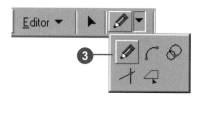

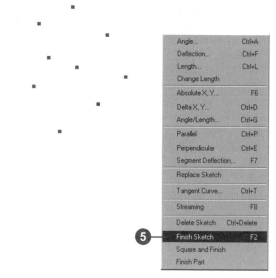

Now, when you click one part of the multipoint feature to select it, all points are automatically selected because they all belong to one multipoint feature.

All points of the feature are selected.

Creating lines and polygons

You can create lines or polygons by digitizing the vertices that make up the feature. For example, to create a square building, you would digitize the four corners. Use any combination of methods for creating vertices or segments. ▶

Tip

Deleting a vertex

To delete a single vertex from a sketch, center the pointer over the vertex until the pointer changes. Right-click, then click Delete Vertex.

Tip

Deleting the sketch

To delete the entire sketch of the feature you're creating, position the pointer over any part of the sketch, right-click, and click Delete Sketch. You can also delete a sketch by pressing Ctrl + Num Del.

Tip

Shortcut for finishing the sketch

You can double-click on the last vertex of the feature to finish the sketch. Or, press F2 when you've finished creating the sketch.

Creating a line or polygon feature by digitizing

1. Click the Current Task dropdown arrow and click Create New Feature.

2. Click the Target layer dropdown arrow and click a line or polygon layer.

3. Click the tool palette dropdown arrow and click the Sketch tool.

4. Click on the map to digitize the feature's vertices.

5. When finished, right-click anywhere on the map and click Finish Sketch. ▶

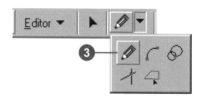

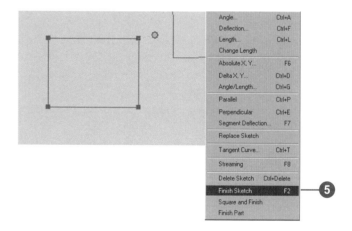

The Square and Finish command on the Sketch tool context menu is a way of completing a polygon. It finishes a polygon by adding two new segments at 90-degree angles. Square and Finish saves you time and ensures precision when creating square-cornered buildings.

ArcMap also provides a way to create a *multipart feature*, a feature that is composed of more than one physical part but only references one set of attributes in the database. For example, the State of Hawaii could be considered a multipart feature. Although composed of many islands, it would be recorded as one feature. A multipart feature can only share vertices, not edges.

Tip

"Undoing" and "redoing" a vertex

You can undo the last vertex you created by clicking the Undo button on the ArcMap Standard toolbar. Click the button again to undo the second-to-last vertex you created, and so on. Click the Redo button if you want to readd the vertex.

The line or polygon is created on your map.

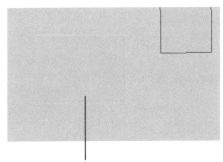

A new feature is created on your map.

Shortcut for finishing a part

You can hold down the Shift key and double-click on the last vertex of a part to finish it.

Shortcut for finishing the sketch

You can double-click on the last vertex of the new feature to finish the sketch.

Streaming

You can also create lines and polygons with the mouse using stream mode digitizing (streaming). For more information, see Chapter 5, 'Using a digitizer'.

Replace sketch

You can add the shape of a line or polygon feature to the sketch by right-clicking over the feature with the Sketch tool and clicking Replace Sketch. The sketch will contain the shape of the feature you clicked over.

Creating a multipart line or polygon

1. Create a line or polygon feature.

2. When finished creating the first part of the feature, right-click anywhere on the map and click Finish Part.

3. Create the next part of the feature.

4. When you have finished the last part of the feature, right-click anywhere on the map and click Finish Sketch. ▶

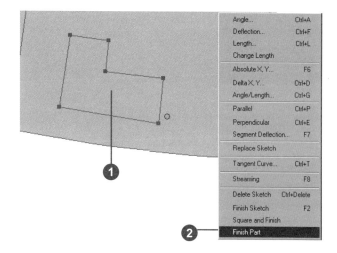

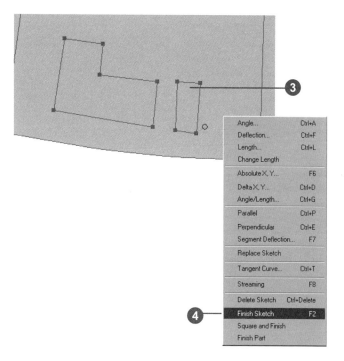

Now when you click one part of the feature to select it, all parts are automatically selected because they all belong to one multipart feature.

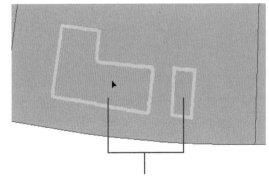

Both parts of the feature are selected.

Squaring a polygon or polyline

1. Click the Current Task dropdown arrow and click Create New Feature.

2. Click the Target layer dropdown arrow and click a polygon or polyline layer.

3. Click the tool palette dropdown arrow and click the Sketch tool.

4. Digitize at least two segments.

5. Right-click anywhere away from the sketch and click Square and Finish. ▶

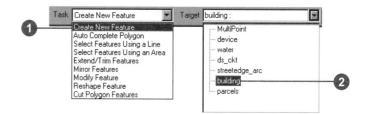

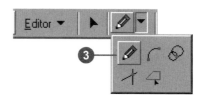

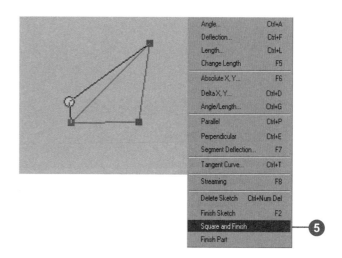

The angles from the first vertex and the last vertex are squared. A new vertex is added, and the sketch is finished where the resulting segments intersect.

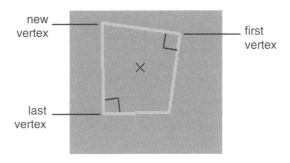

new vertex

first vertex

last vertex

Creating segments using angles and lengths

The edit tools help you create segments at specific angles, measured either using the map coordinate system (Angle) or from the last segment (Deflection).

The Angle command uses east as 0 degrees and measures positive angles counterclockwise. For example, a 90-degree angle represents north and a 180-degree angle represents west.

The Deflection command uses the last segment as 0 degrees and calculates the angle you specify from there. Positive values are calculated in a counterclockwise direction from the existing segment, while negative values are calculated clockwise. You might use Deflection to create the bent end of a water or gas line at a 33-degree angle to a house. ▶

Tip

Shortcut for angle

After clicking the Sketch tool and creating at least one vertex, you can press Ctrl + A to set the angle.

Creating a segment using an angle and a length

1. Click the tool palette dropdown arrow and click the Sketch tool after creating at least one vertex.

2. Right-click away from the sketch and click Angle.

3. Type the angle and press Enter.

   The segment is constrained to the specified angle. ▶

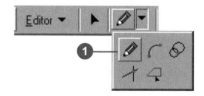

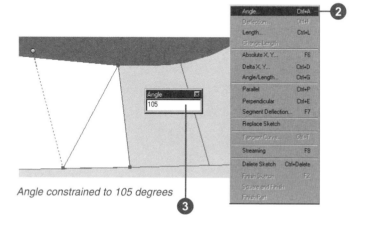

Angle constrained to 105 degrees

Both the Angle and Deflection commands constrain the angle of the segment. For example, if you type "45" as the angle, the segment will be constrained to a 45-degree angle one way and a 225-degree angle the other.

Use the Length command to specify the length of a segment you're creating.

Tip

Shortcut for length

After clicking the Sketch tool and creating at least one vertex, you can press Ctrl + L to set the length.

Tip

Specifying angle and length at the same time

Choose the Angle/Length command from the Sketch tool context menu. After clicking the Sketch tool and creating at least one vertex, you can press Ctrl + G to set the angle and length.

Tip

Changing the length of a segment

If you want to change the length of a segment you have already created, you can use Change Length on the Sketch tool context menu. This undoes the last vertex while keeping the angle constraint.

When using the Sketch tool, you can press F5 to undo the last vertex while keeping the angle constraint.

4. Right-click anywhere on the map and click Length.
5. Type the length and press Enter.

   The vertex that makes the segment the desired angle and length is created.

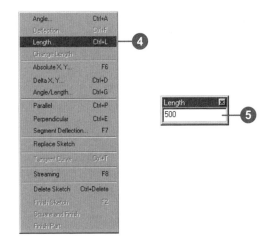

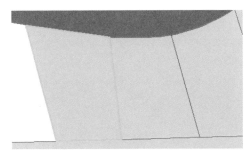

The vertex that makes the segment the desired angle and length is created.

Tip

Shortcut for deflection

After clicking the Sketch tool and creating at least one vertex, you can press Ctrl + F to set the deflection angle.

Creating a segment at an angle from the last segment (deflection)

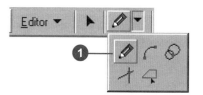

1. Click the tool palette dropdown arrow and click the Sketch tool after creating at least one vertex for the new segment.

2. Right-click away from the sketch.

3. Click Deflection.

4. Type the desired angle from the last segment and press Enter. ▶

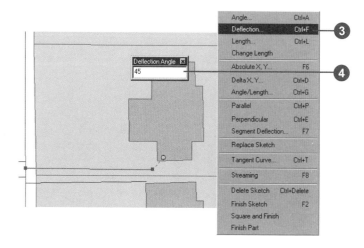

The segment is constrained to the specified angle.

5. Click once to digitize the endpoint of the segment or choose Length from the Sketch tool context menu.

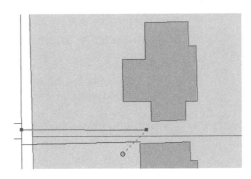

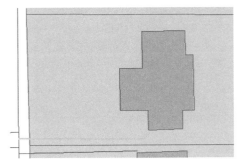

Creating segments using angles from existing segments

Three commands on the Sketch tool context menu—Segment Deflection, Parallel, and Perpendicular—help create segments with angles relative to segments that already exist.

The Segment Deflection command lets you create a segment at an angle relative to any existing segment. While Deflection creates a segment at a specific angle from the last segment in the feature you're creating, Segment Deflection lets you choose which segment you want to work from—it need not be the last segment.

As with the Deflection command, the segment you work ▶

Tip

Shortcut for segment deflection

After clicking the Sketch tool, creating at least one vertex, and positioning the pointer over the segment from which you want to be the specific angle, you can press F7 to set the angle.

Creating a segment at an angle from any other segment

1. Click the tool palette dropdown arrow and click the Sketch tool after creating at least one vertex.

2. Position the pointer over the segment from which you want to create a segment and right-click with the mouse.

3. Click Segment Deflection.

4. Type the desired angle from the segment you chose and press Enter. ▶

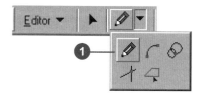

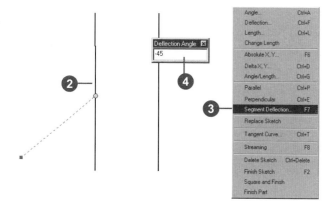

from with Segment Deflection is 0 degrees, and the deflection angle you specify for the new segment is calculated from there. Positive values are calculated in a counterclockwise direction from the existing segment, while negative values are calculated clockwise. The example given shows a cross street created at a -45-degree angle to the existing streets.

The Parallel command on the Sketch tool context menu constrains a segment to be parallel to any segment you choose. For instance, you might use this command to create a gas main line parallel to the street.

The Perpendicular command on the Sketch tool context menu constrains a segment to be perpendicular to an existing segment. You might use this command to place a service line perpendicular to the main line.

Tip

Using only positive values with segment deflection

If you wish to work only with positive angle values, convert negative angles to positive angles by adding 180 to the negative value. For example, a -45-degree angle (measured clockwise) becomes a 135-degree angle (measured counterclockwise).

The segment is constrained to the specified angle.

5. Click once to digitize the endpoint of the segment or choose Length from the Sketch tool context menu.

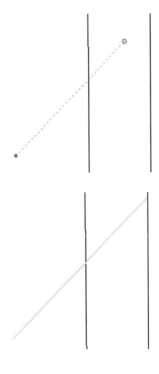

Creating a segment parallel to another segment

1. Click the tool palette dropdown arrow and click the Sketch tool after creating at least one vertex.

2. Position the pointer over the segment to which the new segment will be parallel and right-click.

3. Click Parallel.

   The segment is constrained to be parallel to the specified segment.

4. Click once to digitize the endpoint of the segment or choose Length from the Sketch tool context menu.

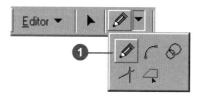

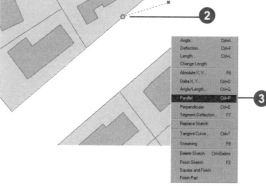

Shortcut for perpendicular

After clicking the Sketch tool, creating at least one vertex, and positioning the pointer over the segment to which the new segment will be perpendicular, you can press Ctrl + E to make the segment perpendicular.

Creating a segment perpendicular to another segment

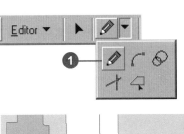

1. Click the tool palette dropdown arrow and click the Sketch tool after creating at least one vertex.

2. Position the pointer over the segment to which the new segment will be perpendicular and right-click with the mouse.

3. Click Perpendicular.

   The segment is constrained to be perpendicular to the specified segment.

4. Click once to digitize the endpoint of the segment or choose Length from the Sketch tool context menu.

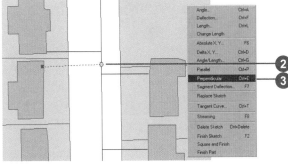

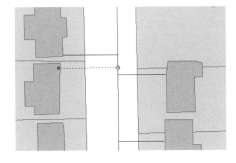

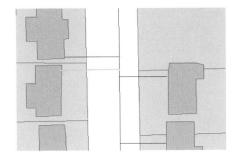

Creating segments that are parametric curves

When creating features, it is often necessary to create a *parametric (true) curve*. Instead of being made of numerous vertices, a parametric curve has only two vertices as endpoints. ArcMap offers two ways to create a segment that is a parametric curve.

First, you can create a parametric curve using the Arc tool. You might use the Arc tool to digitize a cul-de-sac using an aerial photo image as a backdrop.

You can also create a parametric curve using the Tangent Curve command on the Sketch tool context menu. You can use the Tangent Curve command to add a parametric curve to an existing segment. For example, you might use this command to add a curved segment to extend a centerline along a curved road.

When you create a tangent curve, you must specify two parameters for the curve from the following options: arc length, chord, radius, or delta angle. You must also specify whether you want to create ▶

Creating a segment that is a parametric curve using the Arc tool

1. Click the tool palette dropdown arrow and click the Arc tool.

2. Click once to establish the startpoint of the arc.

    A vertex is created.

3. Click once to establish the axis of the arc.

    This is the invisible point through which the curve passes.

4. Click once to establish the endpoint of the arc.

    A segment that is a true curve is created.

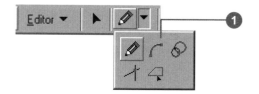

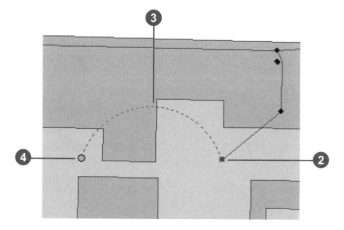

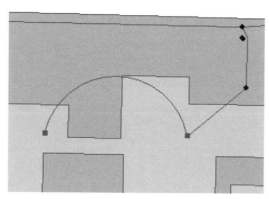

A segment that is a true curve is created.

the curve to the right of the line or to the left of the line, according to the direction in which the line was drawn. The curve is created from the last vertex of the existing segment based on the parameters you defined.

If you choose chord length and radius to construct the curve, there are two possible solutions —the major and minor portions of the circle.

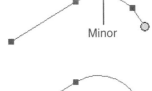

Minor

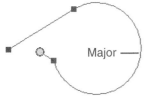

Major

The Minor check box will appear at the bottom of the Tangent Curve dialog box. Check it to construct the minor portion of the circle.

Creating a segment that is a parametric curve using the Tangent Curve command

1. Click the tool palette dropdown arrow and click the Sketch tool after creating at least one segment.

2. Right-click anywhere on the map and click Tangent Curve.

3. Click the dropdown arrows and click two parameters by which you want to define the curve.

4. Type the appropriate values for the parameters (distance in map units for arc length, chord, and radius; degrees for delta angle).

5. Click Left to create the tangent curve to the left of the segment. Click Right to create the curve to the right.

6. Press Enter. ▶

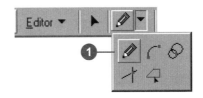

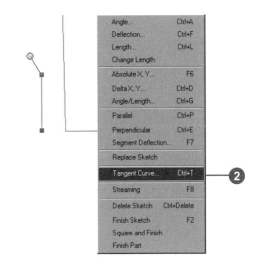

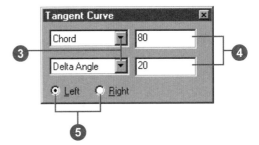

A segment that is a true curve
is created from the last vertex
of the segment according to
the parameters you specified.

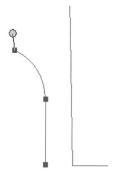

*A parametric curve with a chord length of 80 map units and a delta
angle of 70 degrees is created to the left of the last vertex.*

Creating segments by tracing features

You can create segments by tracing over the segments of selected features using the Trace tool.

Suppose you want to create a new water main that is offset 7 meters from the parcel boundaries. Using the Trace tool, you can create new segments in the sketch that are at the same angle as the selected parcel boundaries yet constructed at an offset value of 7 meters.

Tip

Backing up a trace
If you traced too far or have traced the wrong direction, move the mouse backwards over what you have traced. If you have clicked to stop the trace, click Undo to remove all vertices added during the trace.

Tip

Canceling a trace
A quick way to cancel a trace is to Press the Esc key.

Tip

Finish the sketch
When you are finished tracing, you can double-click to finish the sketch.

Creating segments by tracing features

1. Click the Edit tool.

2. Select the features that you want to trace.

3. Click the tool palette dropdown arrow and click the Trace tool.

4. Press O to set an offset value. Type an offset value and press Enter. Setting an offset value is optional. If you want to trace directly on top of existing features, enter a value of 0.

5. Click to start tracing.

6. Click to stop tracing.

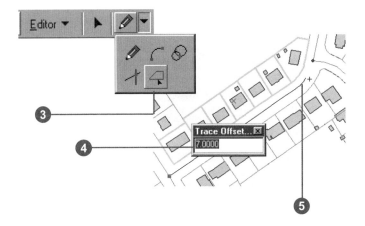

The Snapping Environment window

You can keep the window open as you work—any changes in settings are effective immediately. Click the Close button when you are finished.

The layers in your map document are listed here. Set the snapping priority—the order in which snapping will occur by layer—by dragging the layer names to new locations.

The bottom portion of the window shows snapping properties that work with a sketch.

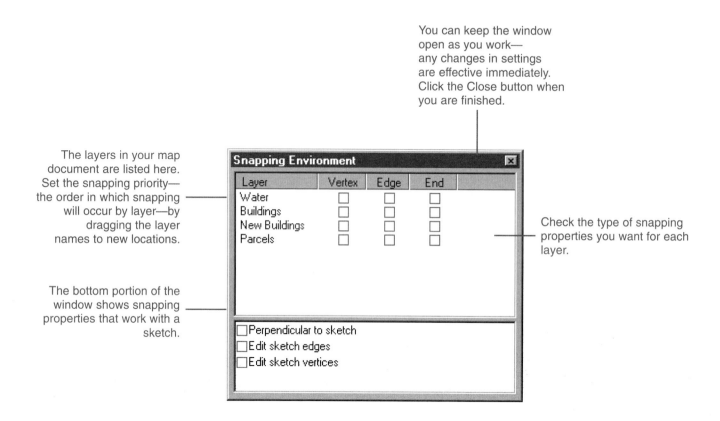

Check the type of snapping properties you want for each layer.

Types of snapping properties

When you use the snapping environment to create or place a new feature in an exact location relative to other features, you must choose to which part of existing features—vertex, edge, or endpoint—you want your feature to snap. These choices are called layer *snapping properties*. You can also specify snapping properties for the edit sketch itself; these are called sketch snapping properties. You can set both types of snapping properties using the Snapping Environment window. The following table briefly explains each of the layer snapping and sketch snapping properties.

Layer snapping properties		Sketch snapping properties	
Vertex	Snaps to each vertex of the features in that layer.	Perpendicular to sketch	Lets you create a segment that will be perpendicular to the previous.
Edge	Snaps to the entire outline (both segments and vertices) of each feature in that layer.	Edit sketch vertices	Snaps to the vertices of the sketch.
Endpoint	Snaps to the first vertex and the last vertex in a line feature.	Edit sketch edges	Snaps to the entire outline (both segments and vertices) of the sketch.

Using the snapping environment

The *snapping environment* can help you establish exact locations in relation to other features. Suppose you're creating a new segment of primary that begins from an existing transformer; you want to ensure that the vertex of the primary connects precisely to the transformer.

The snapping environment makes this type of task accurate and easy. Setting the snapping environment involves setting a snapping tolerance, snapping properties, and a snapping priority.

The *snapping tolerance* is the distance within which the pointer or a feature is snapped to another location. If the location being snapped to (vertex, edge, or endpoint) is within the distance you set, the pointer automatically snaps (jumps) to the location. ▶

Tip

Viewing the snapping tolerance

To see the current snapping tolerance, hold down the T key while using the Sketch tool.

Setting the snapping tolerance

1. Click Editor and click Options.

2. Click the General tab.

3. Click the Snapping tolerance dropdown arrow and click the type of measurement unit you want to use for the snapping tolerance—pixels or map units.

4. Type the desired number of measurement units in the Snapping tolerance text box.

5. Click OK.

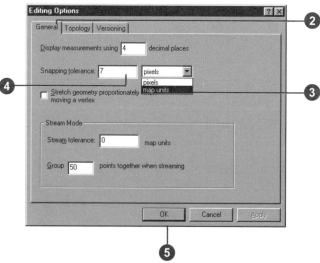

The circle around the pointer in the graphics below represents the snapping tolerance. When the location being snapped to (orange point) is outside the snapping tolerance, the snapping location (blue dot) stays with the pointer (top graphic). When the location being snapped to is inside the snapping tolerance, the snapping location moves away from the pointer and snaps to the target location (bottom graphic).

You can choose the part of the feature—vertex, edge, or endpoint—to which you want your new feature to snap by setting the layer *snapping properties*. For example, if you want your new feature—a segment of primary—to snap to the vertex of an existing transformer in the transformers layer, you would check the box under Vertex and next to the transformers layer in the Snapping Environment window. When ▶

Setting snapping properties

1. Click Editor and click Snapping.

   The Snapping Environment window appears.

2. Check the snapping properties you want.

   The snapping properties are effective as soon as they are checked or unchecked.

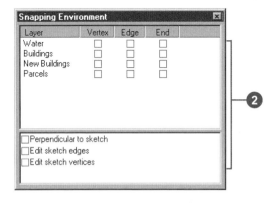

the pointer comes within the snapping tolerance of the transformer, the first vertex of the primary snaps to the vertex of the transformer.

You can also set the *snapping priority* for layers on your map. The order of layers listed in the Snapping Environment window determines the order in which snapping will occur. Snapping occurs first in the layer at the top of the list and then in each consecutive layer down the list. You can easily change the snapping priority by dragging the layer names to new locations.

Tip

Sketch snapping properties

You can set snapping properties that apply specifically to an edit sketch in the Snapping Environment dialog box as well; these are located at the bottom of the Snapping Environment window. For more information, see 'Types of snapping properties' in this chapter.

Setting the snapping priority

1. Click Editor and click Snapping.

   The Snapping Environment window appears.

2. Click and drag the layer names to arrange them in the order in which you want snapping to occur. (The first layer in the list will be snapped to first.)

   The snapping priorities you set are effective immediately.

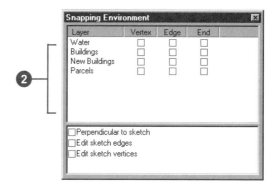

Using a digitizer

5

Digitizing is the process of converting features on a paper map into digital format. To digitize a map, you use a *digitizing tablet* connected to your computer to trace over the features that interest you. The x,y coordinates of these features are automatically recorded and stored as spatial data.

Digitizing with a digitizing tablet offers another way, besides digitizing "freehand", to create and edit spatial data. You can convert features from almost any paper map into digital features. You can use a *digitizer* in conjunction with the tools in ArcMap to create new features or edit existing features on a digital map.

You may want to digitize features into a new layer and add the layer to an existing map document, or you may want to create a completely new set of layers for an area for which no digital data is available. You can also use a digitizer to update an existing layer on your digital map.

Chapter 4, 'Creating new features', introduced you to the Sketch tool and other useful editing tools in ArcMap and discussed how these are used to digitize features freehand. This chapter will teach you the fundamentals of editing features in ArcMap using a digitizer. You may want to read Chapter 4 first to get an understanding of editing before reading this chapter.

Setting up your digitizing tablet and preparing your paper map

Before you can start digitizing, you must set up your digitizing tablet and prepare your paper map. This can be done after you have installed the digitizer driver software.

Installing the driver software and configuring puck buttons

To use a digitizing tablet with ArcInfo, it must have WinTab™-compliant digitizer driver software. To find out if a WinTab-compliant driver is available for your digitizer, see the documentation that came with the tablet or contact the manufacturer.

After installing the driver software, use the WinTab manager setup program to configure the buttons on your *digitizer puck.* (You may have to turn on your digitizer and reboot your machine before you can use the setup program.) One puck button should be configured to perform a left mouse click to digitize point features and vertices; another button should be configured to perform a left double-click to finish digitizing line or polygon features.

With any development programming language, you can configure additional buttons to run specific ArcMap commands—such as the Zoom In or Sketch tools—normally accessed through toolbar buttons and menus. *Exploring ArcObjects* contains sample Visual Basic® for Applications (VBA) code that you can use to run a variety of ArcMap commands from the digitizer puck.

Preparing the map

After you have set up your digitizing tablet and configured the puck buttons, you can prepare your paper map for digitizing. Your map should ideally be reliable, up-to-date, flat, and not torn or folded. Paper expands or shrinks according to the weather. To minimize distortion in digitizing, experienced digitizers often copy paper maps to a more stable material such as Mylar®.

If you know what coordinate system (projection) your paper map is in, you should set the same projection for the layer you're digitizing into. If you are digitizing features into an existing feature layer, you must ensure that your paper map and digital layer share the same coordinate system. For more information on specifying a coordinate system in ArcMap, see Chapter 4, 'Creating maps', in *Using ArcMap.*

Establishing control points on your paper map

Before you can begin digitizing from your paper map, you must first establish *control points* that you will later use to register the map to the geographic space in ArcMap. If your map has a grid or a set of known ground points, you can use these as your control points. If not, you should choose between four and ten distinctive locations such as road intersections and mark them on your map with a pencil. Give each location a unique number and write down its actual ground coordinates.

Once you've identified at least four well-placed control points, you can place your map on the tablet and attach it with masking tape. You don't have to align the map precisely on your tablet; ArcMap corrects any alignment problems when you register the map and displays such adjustments in the error report.

The error report includes two different error calculations: a point-by-point error and a root mean square (RMS) error. The point-by-point error represents the distance deviation between the transformation of each input control point and the corresponding point in map coordinates. The RMS error is an average of those deviations. ArcMap reports the point-by-point error in current map units. The RMS error is reported in both current map units and digitizer inches. If the RMS error is too high, you can reregister the appropriate control points. To maintain highly accurate data, the RMS error should be kept under 0.004 digitizer inches. For less accurate data, the value can be as high as 0.008 digitizer inches.

Registering your paper map

Before you can start digitizing, you must register your paper map into real-world coordinates. This allows you to digitize features directly in geographic space.

Registering your map involves recording the ground coordinates for the control points you identified while preparing your map. These are recorded using the Digitizer tab of the Editing Options dialog box. You must first use the digitizer puck to digitize the control points on the paper map; with the puck over each control point on the map, press the button you configured to perform a left mouse click. You must then type the actual ground coordinates for each control point.

When registering your map, you have the option of saving the ground coordinates you entered for later use—for example, if you want to reregister your map or register another map that uses the same control points. These ▶

See Also

For information on configuring puck buttons and establishing control points, see 'Setting up your digitizing tablet and preparing your paper map' in this chapter.

Registering your map for the first time

1. After adding a layer to your map, click the Editor menu and click Start Editing.

2. Click Editor and click Options.

3. Click the Digitizer tab.

4. With the digitizer puck, digitize the control points you established earlier on your paper map.

   A record appears in the X Digitizer and Y Digitizer columns for each control point you digitized.

5. Type the actual ground coordinates for each control point in the X Map and Y Map fields.

   An error in map units is displayed at each control point. An RMS error is displayed in map units and in digitizer inches.

6. Click OK to register the map and close the Editing Options dialog box.

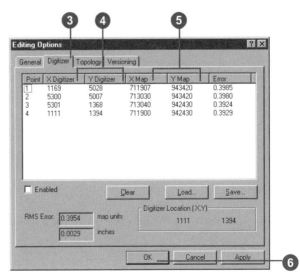

ground coordinates are stored in tic text files.

After you've entered the ground coordinates, ArcMap displays an error at each control point as well as an RMS error. If the RMS error is too high—greater than 0.004 digitizer inches for highly accurate data or greater than 0.008 digitizer inches for less accurate data—you can register the appropriate control points again. For more information on errors, see 'Setting up your digitizing tablet and preparing your paper map' in this chapter.

Tip

Missing Digitizer tab
If you installed ArcInfo before installing your digitizer, the Digitizer tab may be missing from the Editing Options dialog box. To add the tab, you must register the digitizer.dll file. Go to the DOS prompt, type "cd" followed by the path to the directory where you installed ArcInfo (%ARCHOME%\bin), and type "regsvr32 digitizer.dll". When you restart ArcMap, the Editing Options dialog box will have the Digitizer tab.

Saving new ground coordinates

1. Follow steps 1 through 5 for registering your map for the first time.

2. Click Save.

3. Navigate to the directory in which you want to save the coordinates and type a filename.

4. Click Save.

5. Click OK.

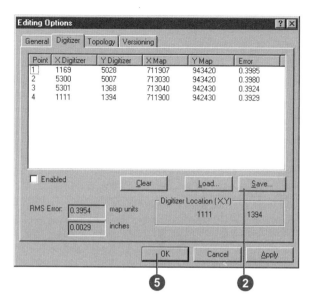

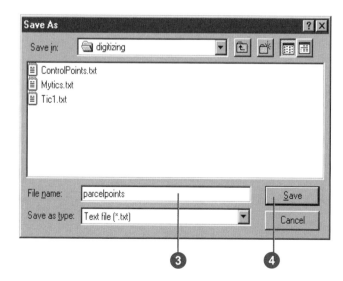

Tip

Removing records

If you want to remove all the ground coordinate records and start over, click Clear on the Digitizer tab. To remove an individual record, click the number in the Point column corresponding to the coordinates you want to remove and press the Delete key.

Tip

Adding records

If you want to add additional control points after entering a few, click below the last record with the mouse and digitize the new points with the digitizer puck.

Tip

Digitizer location

The Digitizer tab also displays the current x,y location of the digitizer puck on the tablet. The coordinates change as you move the puck along the tablet surface. This helps orient you to the location you're digitizing.

Registering your map using existing tic files or saved coordinates

1. After adding a layer to your map, click the Editor menu and click Start Editing.

2. Click the Editor menu and click Options.

3. Click the Digitizer tab.

4. Click Load.

5. Navigate to the file you want to use.

6. Click Open. ▶

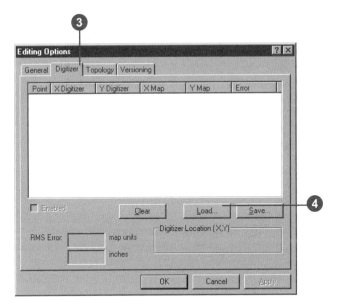

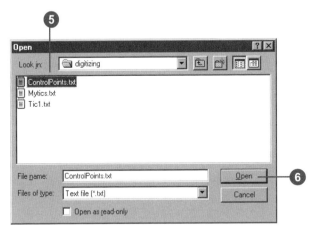

Digitizing accuracy

Always register your map at the start of each digitizing session, even if this means registering the same map more than once. Your paper map might shift between sessions; reregistering helps ensure that your digitizing is accurate.

The ground coordinates appear under the X Map and Y Map fields.

7. Click the first record and digitize the first control point with the digitizer puck.

8. Digitize each of the other control points.

   The digitized coordinates appear in the X Digitizer and Y Digitizer columns. An error is displayed for each control point, and an RMS error is displayed in map units and in digitizer inches.

9. Click OK to register the map.

The ground coordinates are displayed.

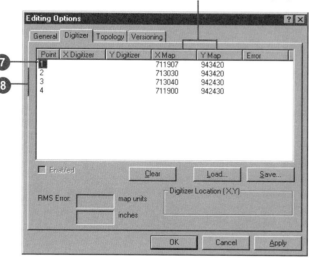

The digitized coordinates are displayed.

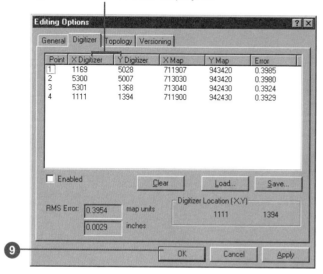

Creating features using a digitizer

It's easy to digitize features in ArcMap. You can digitize features into a new map layer or edit an existing layer.

Digitizing modes

Digitizing tablets generally operate in two modes: digitizing (absolute) mode and mouse (relative) mode.

In *digitizing mode*, the location of the tablet is mapped to a specific location on the screen. In other words, moving the digitizer puck on the tablet surface causes the screen pointer to move to precisely the same position. When you are in digitizing mode, you can only digitize features; you can't choose buttons, menu commands, or tools from the ArcMap user interface because the screen pointer is locked to the drawing area.

In *mouse mode*, the digitizer puck behaves just like a mouse; there is no correlation between the position of the screen pointer and the surface of the digitizing tablet, but you can choose interface elements with the pointer.

ArcMap lets you switch between digitizing and mouse modes using the Editing Options dialog box. This means you can use the digitizer puck to both digitize features and access user interface choices (as a substitute to the mouse) as you digitize.

Whether your digitizer is in mouse mode or digitizing mode, you can still use your mouse at any time to choose interface elements.

Two ways to digitize features on a paper map

You can digitize features on a paper map in two ways: using point mode digitizing or stream mode digitizing (streaming). You can switch back and forth between the two modes as you digitize by pressing F8.

Digitizing by point

When you start a digitizing session, the default is point mode. With *point mode digitizing*, you convert a feature on a paper map by digitizing a series of precise points, or vertices. ArcMap then connects the vertices to create a digital feature. You would use point mode when precise digitizing is required—for example, when digitizing a perfectly straight line.

Digitizing using stream mode

Stream mode digitizing (streaming) provides a quick and easy way to capture features on a paper map when you don't require as much precision—for example, to digitize rivers, streams, and contour lines. With stream mode, you create the first vertex of the feature and trace over the rest of the feature with the digitizer puck. When you're finished tracing, you use the puck to complete the feature.

As you stream, ArcMap automatically adds vertices at an interval you specify; this interval, expressed in current map units, is called the stream tolerance. You can change the stream tolerance at any time, even while you're in the process of digitizing a feature.

You can also digitize using stream mode when you create features "freehand" with the sketch construction tools. You can digitize in stream mode with the Sketch tool, for example, in the same way you do from a paper map. The only difference is that you use the mouse pointer to digitize freehand.

Digitizing features in point mode

Point mode digitizing works the same way with a digitizer as with "freehand" digitizing with the Sketch tool; the only difference is that with the digitizer you're converting a feature from a paper map using a digitizer puck instead of a mouse.

Point mode digitizing involves converting point, line, and polygon features from a paper map by digitizing a series of precise points, or vertices. You digitize each vertex by pressing the puck button you configured to perform a left mouse click. To finish the feature, press the puck button you configured to perform a left double-click. ArcMap connects the vertices to create a digital feature.

Before you begin digitizing, you must set the digitizer to work in digitizing mode, rather than in mouse mode; this constrains the screen pointer to the digitizing area. When the puck is in ▶

See Also

For information on configuring puck buttons and establishing control points, see 'Setting up your digitizing tablet and preparing your paper map' in this chapter.

1. Click Editor and click Options.
2. Click the Digitizer tab.
3. Check Enabled to use the puck in digitizing mode.
4. Click OK. ▶

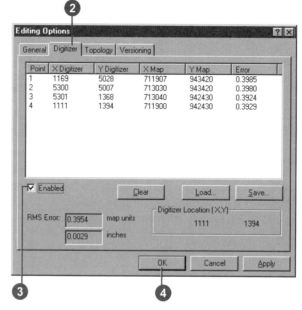

digitizing mode, you must use your mouse to choose items in the ArcMap interface—unless you have used VBA, or another development programming language, to configure additional puck buttons to run specific ArcMap commands.

Tip

Snapping

To help you digitize features in a precise location on an existing layer, you can use the snapping environment. For information on snapping, see Chapter 4, 'Creating new features'.

Tip

Deleting vertices

Click the Undo button on the ArcMap Standard toolbar to delete a vertex as you digitize.

See Also

For information on creating features by digitizing freehand with the sketch creation tools, see Chapter 4, 'Creating new features'.

See Also

For information on configuring puck buttons with programming code, see 'Setting up your digitizing tablet and preparing your paper map' in this chapter.

5. Click the tool palette dropdown arrow and click the Sketch tool.

6. With the digitizer puck, digitize the first vertex of the feature.

7. Trace the puck over the feature on the paper map, creating as many vertices as you need.

8. Finish the feature by pressing the appropriate puck button.

   The feature is created.

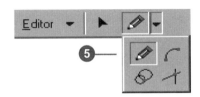

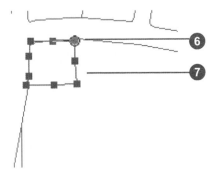

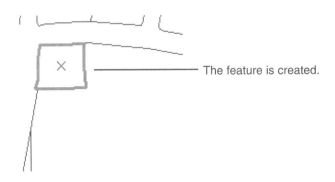

The feature is created.

Digitizing features in stream mode

When you digitize line or polygon features from a paper map in stream mode (streaming), you create the first vertex of the feature by pressing the digitizer puck button you configured to perform a left mouse click. You then trace over the rest of the feature with the digitizer puck. When you're finished tracing, press the puck button you configured to perform a left double-click to complete the feature.

Before starting to digitize in stream mode, you must set the *stream tolerance*—the interval at which ArcMap adds vertices along the feature you're digitizing. Because the default stream tolerance is 0, you must enter a tolerance value before you start digitizing or the vertices will join together or overlap each other. You can change the stream tolerance any time in the digitizing process.

You must also specify the number of streaming vertices you want to group together. The number you set tells ArcMap how many vertices to delete when you click the Undo button. For example, if you set this ►

Setting the stream tolerance

1. Click the Editor menu and click Start Editing.
2. Click the Editor menu and click Options.
3. Click the General tab.
4. Type the stream tolerance (in map units) in the Stream tolerance text box.
5. Click OK.

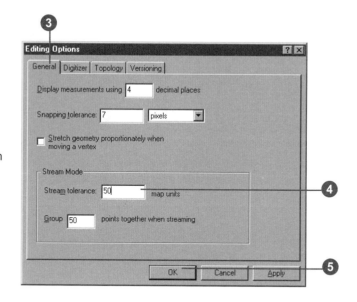

Setting the number of vertices to be grouped

1. Click the Editor menu and click Options.
2. Click the General tab.
3. Type the number of vertices you want to group together.
4. Click OK.

   Now when you click the Undo button while digitizing in stream mode, the number of vertices you specified are deleted.

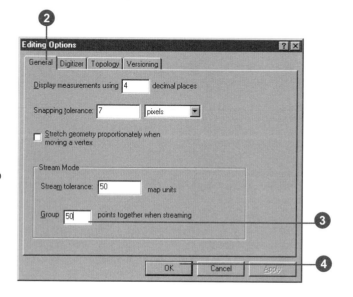

number to 20 and click the Undo button while you're digitizing a feature, ArcMap deletes the last 20 digitized vertices from your feature.

To begin digitizing in stream mode, you must choose Streaming from the Sketch tool context menu. You can switch back to point mode at any time by pressing F8; press F8 again to switch to stream mode again.

Before streaming, remember to set the digitizer to work in digitizing mode rather than in mouse mode; this constrains the screen pointer to the digitizing area.

Tip

Snapping
To help you digitize features in a precise location on an existing layer, you can use the snapping environment. For information on snapping, see Chapter 4, 'Creating new features'.

Digitizing a feature in stream mode

1. Click the Editor menu and click Options.

2. Follow steps 3 and 4 for setting the stream tolerance.

3. Follow step 3 for setting the number of vertices to be grouped.

4. Click the Digitizer tab.

5. Check Enabled to use the puck in digitizing mode.

6. Click OK.

7. Click the tool palette dropdown arrow and click the Sketch tool. ▶

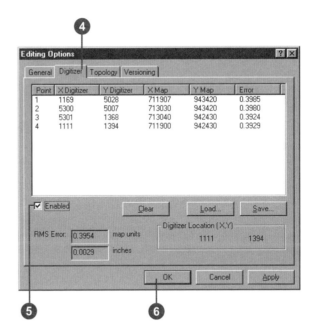

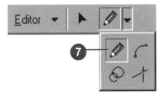

Choosing interface elements while streaming

When you're in the process of digitizing a feature in stream mode and want to interact with the ArcMap interface using your mouse—for example, to change the stream tolerance or undo an action—you must first switch back to point mode by pressing F8. After you have finished interacting with the interface, you can resume streaming by pressing F8 again.

Configuring a puck button for streaming

Instead of choosing Streaming from the context menu, you can configure one of your puck buttons using any development programming language, such as C++ or VBA, to activate stream mode digitizing. To learn more about configuring your puck buttons and customization in general, see Exploring ArcObjects.

8. With the mouse pointer, right-click anywhere on the map and click Streaming.

9. With the digitizer puck, digitize the first vertex of the line or polygon feature.

10. Trace the puck over the feature on the paper map.

    ArcMap creates vertices at the stream tolerance you specified.

11. Finish the feature by pressing the appropriate puck button.

    The feature is created.

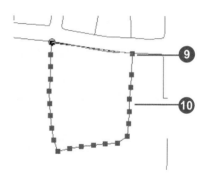

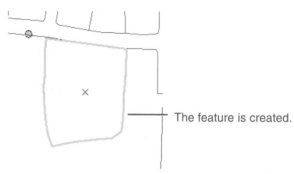

The feature is created.

Creating features from other features

6

In spatial data editing, many new features can be created using the shapes of other features. ArcMap has many tools you can use to create new features based on features already in your database.

For example, you can construct a line that is a parallel copy of an existing line to create a centerline on a street. You can create a buffer around a point, line, or polygon feature to show a specific area, such as a floodplain around a river. You can also create a new feature by combining or intersecting existing features or even create a mirror image of a feature or set of features.

In this chapter, you'll learn how easy it is to perform these tasks using various tools in ArcMap.

Copying a line at a specific interval

The Copy Parallel command copies a line parallel to an existing feature at a distance you specify. If you give a distance that is positive, the line is copied to the right side of the original feature. A negative distance value copies the line to the left.

You might use the Copy Parallel command to create a street centerline or to create a gas line that runs parallel to a road.

1. Click the Edit tool.

2. Click the line you want to copy.

3. Click the Target layer dropdown arrow and click the layer to which you want the new line to belong. ▶

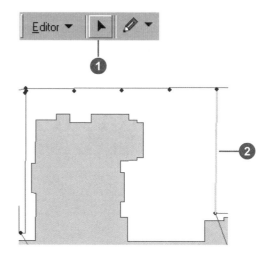

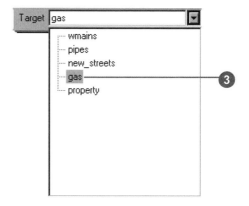

4. Click Editor and click Copy Parallel.

5. Type the distance (in map units) from the original feature where you want to copy the line and press Enter.

   A parallel copy of the line is created at the specified distance.

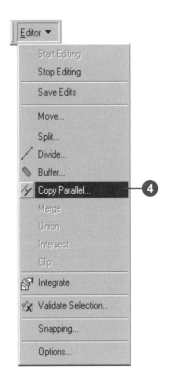

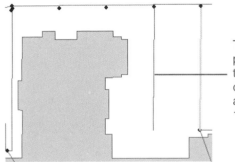

The line is copied parallel and to the left of the original feature at a distance of 150 map units.

Creating a buffer around a feature

You can create a buffer around a feature using the Buffer command. For instance, you might use Buffer to show the area around a well that's contaminated or to represent a floodplain around a river.

You can buffer more than one feature at a time, but a separate buffer will be created around each feature.

1. Click the Edit tool.

2. Click the feature or features around which you want to create a buffer.

3. Click the Target layer dropdown arrow and click the layer with the type of features you want the buffer to be. (This can only be a line or polygon layer.) ▶

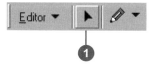

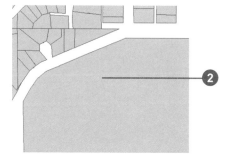

4. Click Editor and click Buffer.

5. Type the distance (in map units) from the feature around which you want to create the buffer and press Enter.

   A buffer is created at the specified distance.

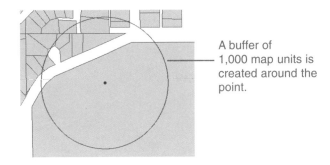

A buffer of 1,000 map units is created around the point.

Creating a mirror image of a feature

The Mirror task creates a mirror image of selected features on the other side of a line you create. You might use the Mirror task to create houses in a housing development where houses are mirror images of the ones on the opposite side of the street.

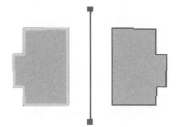

Also, as shown in the example, the Mirror task provides an easy way to add gas services to parcels that mirror the services on the other side of the street.

Tip

Other ways to construct a line

You can also use the Distance–Distance and Intersection tools to create the endpoints of the line. For more information, see 'Creating point features and vertices' in Chapter 4.

1. Click the Edit tool.

2. Click the feature or features that you want to mirror.

3. Click the Current Task dropdown arrow and click Mirror Features.

4. Click the tool palette dropdown arrow and click the Sketch tool.

5. Construct a line by clicking once on the startpoint and once on the endpoint. ▶

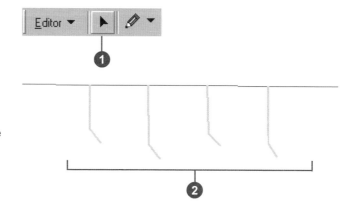

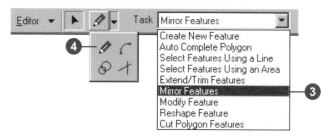

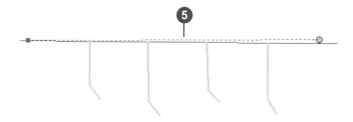

After you digitize the end-point, a mirror image of the feature or features is created.

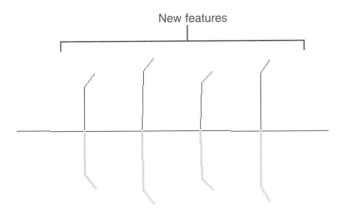

New features

Merging features from the same layer into one feature

The Merge command combines features from the same layer into one feature. The features must be part of a line or polygon layer. You could use the Merge command to combine two parcels into one.

You might also want to merge nonadjacent features to create a multipart feature. For example, you could merge the individual islands that make up Hawaii to create a multipart polygon feature.

When you merge features in a geodatabase, the original features are removed and the new feature's attributes are copied from the feature that was selected first. If you merge coverage or shapefile features, the attributes of the feature with the lowest ID number (the oldest feature) are used.

1. Click the Edit tool.

2. Click the features that you want to merge.

   (The features must be from the same layer, either a line or polygon layer.)

3. Click the Target layer dropdown arrow and click the layer to which you want the new feature to belong. ►

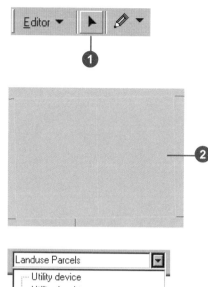

4. Click Editor and click Merge.

   The selected features are merged into one.

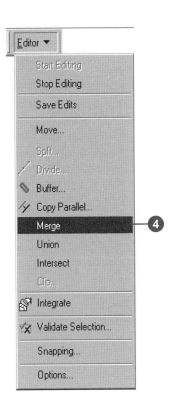

Parcels are merged into one.

Combining features from different layers into one feature

The Union command lets you combine features from different layers into one feature while maintaining the original features and attributes. You might use this command to create a sales territory from several ZIP Codes.

You can also create a multipart feature using the Union command by combining nonadjacent features from different layers. For example, suppose you want to create a sedimentary rock polygon in a new rock classification layer given selected clay and quartz polygons in an existing rock composite layer. You would use the Union command to combine the clay and quartz features to create a new, multipart sedimentary rock feature in the rock classification layer.

When you use the Union command, the features you combine must be from layers of the same type—line or polygon. The new feature is created in the current layer with no attribute values.

1. Click the Edit tool.

2. Click the features that you want to combine into one.

   (The features may be from different layers, although they must be the same layer type—line or polygon.)

3. Click the Target layer dropdown arrow and click the layer to which you want the new feature to belong. ▶

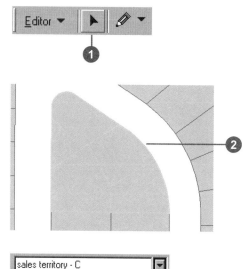

4. Click Editor and click Union.

   The selected features are combined into one.

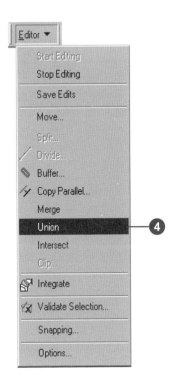

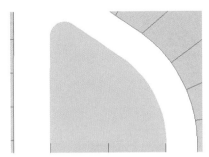

ZIP Codes are combined into one sales territory.

Creating a feature from features with common areas

The Intersect command creates a new feature from the area where features overlap. For instance, you might create a new sales territory out of overlapping trade areas.

You can find the intersection between features of different layers, but the layers must be of the same type—line or polygon. The original features are maintained, and the new feature is created in the current layer with no attribute values. You must enter attribute values for the new feature yourself.

1. Click the Edit tool.

2. Click the features from whose intersection you want to create a new feature.

   (The features may be from different layers, although they must be the same layer type—line or polygon.)

3. Click the Target layer dropdown arrow and click the layer to which you want the new feature to belong.

   (The layer must be of the same type as the selected features—line or polygon.) ►

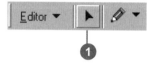

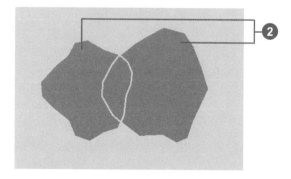

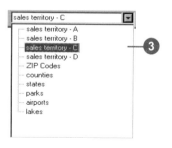

4. Click Editor and click Intersect.

A new feature is created from the areas in common between all selected features.

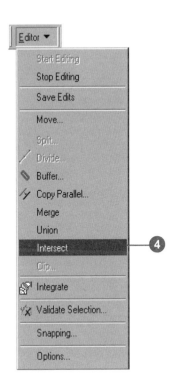

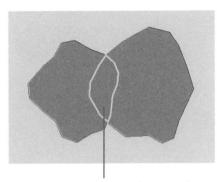

A single sales territory is created from the areas in common between two other sales territories.

Editing existing features

7

This chapter shows you how to modify features that already exist in your database. Suppose you need to change the shape of a parcel to accommodate a newly added cul-de-sac—you can use the Reshape Feature task to modify the parcel to the proper shape. Suppose the street you've digitized doesn't intersect with the correct cross street—you can use the Extend task to extend the line to the correct location. If you need to divide a parcel, you can use the Cut Polygon Feature task to cut the feature into two.

These are just a few examples of how easy it is to modify features while editing in ArcMap. The editing tools, commands, and tasks provide a variety of ways to make changes to existing features.

Splitting a line or polygon

Using the editing tools, you can easily split line and polygon features.

To manually split one line into two, use the Split tool. The line is split at the location where you clicked with the mouse. The attributes of the original line are copied to each of the new lines. In the example shown, the Split tool is used to divide a street centerline into two features in anticipation of a new centerline being added between the parcels.

You can also split a line into two using the Split command on the Editor menu. Use the Split command when you know the distance at which you want to split the line, measured from either the first or last vertex. You can also use this command when you want to split a line at a certain percentage of the original length. You might use the Split ►

You might use the Split ►

Tip

Using snapping to split a line

If you want to use the Split tool to split a line at a specific vertex, use the snapping environment to snap the pointer precisely to the vertex. For more information on snapping, see Chapter 4, 'Creating new features'.

Splitting a line manually

1. Click the Edit tool.
2. Click the line you want to split.
3. Click the Split tool.
4. Click the spot on the line where you want it to split.

   The line is split into two features.

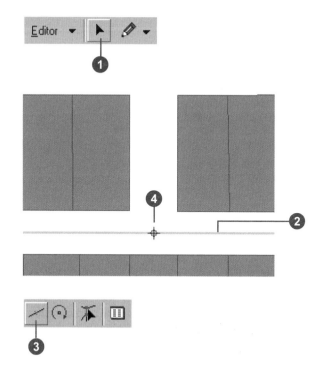

command to split a power line at a known distance along the line when you want to add an electrical pole that requires its own service.

The Split dialog box displays the length of the original feature in current map units to help you split it accurately. When you split the line using the Split command, the attributes of the original line are copied to each of the new lines.

To split one polygon into two, use the Cut Polygon Features task. The polygon is split according to a line sketch you create. The attributes of the original feature are copied to each of the new features.

Splitting a line at a specified distance or percentage

1. Click the Edit tool.

2. Click the line you want to split.

3. Click Editor and click Split. ▶

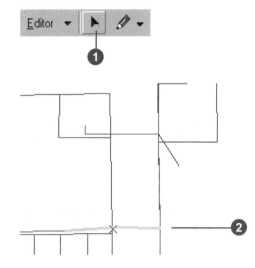

4. Click the first Split option to
   split the feature at a certain
   distance.

   Click the second Split option
   to split the feature at a certain
   percentage of the whole.

5. Type a distance or percentage, as desired.

6. Click Forward if you want to
   split the feature starting from
   the first vertex.

   Click Reverse if you want to
   split the feature starting from
   the last vertex.

7. Click OK.

   The line is split into two
   features according to the
   parameters you specified.

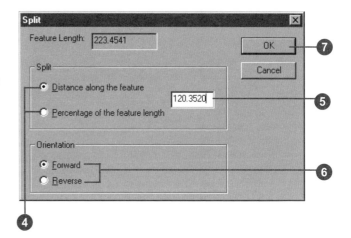

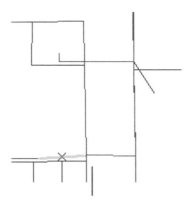

The line is split into two
according to the distance and
orientation you specified.

Tip

Cutting a polygon shape out of a polygon

You can use the Cut Polygon Features task to create a sketch that acts like a "cookie cutter", splitting the polygon in two. Simply create a line sketch that closes in on itself by double-clicking precisely on the first vertex of the sketch to finish it.

Tip

Other ways to construct a sketch

You can also use the Distance–Distance tool, the Arc tool, or the Intersection tool to create a sketch. For more information, see Chapter 4, 'Creating new features'.

Splitting a polygon

1. Click the Edit tool.

2. Click the polygon you want to split.

3. Click the Current Task dropdown arrow and click Cut Polygon Features.

4. Click the tool palette dropdown arrow and click the Sketch tool.

5. Construct a line or polygon sketch that cuts the original polygon as desired.

6. Right-click anywhere on the map and click Finish Sketch.

   The polygon is split into two features.

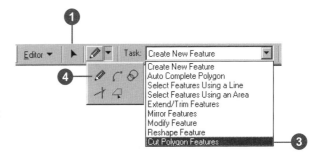

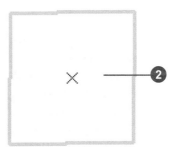

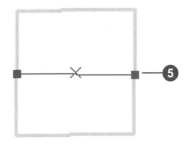

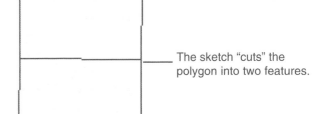

The sketch "cuts" the polygon into two features.

Trimming a line

The Trim command on the Sketch context menu reduces the length of a line, trimming a distance you specify from the last vertex.

The Trim task in the Current Task dropdown list also trims lines, but instead of trimming them a given distance, the Trim task uses a *sketch* you draw. ▶

Tip
Shortcut for modifying features
Instead of using the Modify Feature task to change a feature to its sketch, you can click the Edit tool and double-click the feature you want to modify.

Tip
Trimming from the first vertex of a line
You can trim a line from the first vertex instead of the last. See 'Flipping a line' in this chapter.

Tip
Shortcuts for finishing a sketch
When you're finished modifying a sketch, you can press F2 to finish it. Simply selecting another feature with the Edit tool will also finish the sketch.

Trimming a specific length from the last point

1. Click the Current Task dropdown arrow and click Modify Feature.

2. Click the Edit tool.

3. Click the line that you want to trim.

   The line appears as a sketch with vertices.

4. Right-click over any part of the line and click Trim.

5. Type the length you want to trim from the line (beginning at the last vertex marked in red) and press Enter.

   The line is trimmed.

6. When finished modifying the line, right-click over any part of the sketch and click Finish Sketch.

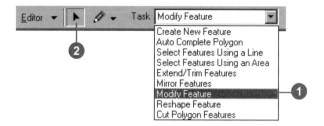

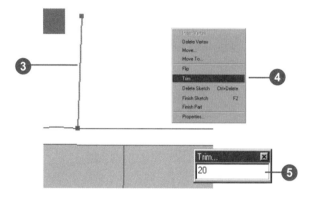

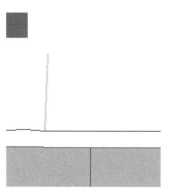

20 map units are trimmed from the original line.

This is useful if you don't know the exact distance you want to trim but have a physical boundary where the features should end or begin.

Suppose your database has some roads that should end at the coastline, but overshoot it instead. Using the Trim task, you can draw a line sketch on top of the coastline and the lines will be trimmed where you have drawn the sketch.

Portions of the lines that are on the right side of the sketch are trimmed. The right side of the sketch is based on the direction in which the sketch was drawn. Imagine riding a bicycle along the sketch in the direction ▶

Tip

Other ways to construct a sketch

You can also use the Distance–Distance tool, the Arc tool, or the Intersection tool to create a sketch. For more information, see Chapter 4, 'Creating new features'.

Tip

Shortcuts for finishing a sketch

You can double-click on the last vertex of a sketch to finish it. You can also press F2.

Trimming based on a line you draw

1. Click the Current Task dropdown arrow and click Extend/Trim Features.

2. Click the Edit tool.

3. Click the line or lines you want to trim.

4. Click the tool palette dropdown arrow and click the Sketch tool.

5. Construct a line that trims the selected line or lines as desired.

6. Right-click anywhere on the map and click Finish Sketch. ▶

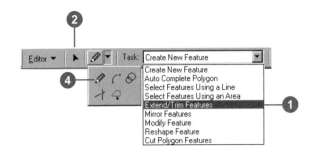

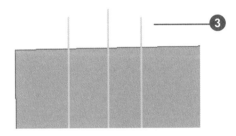

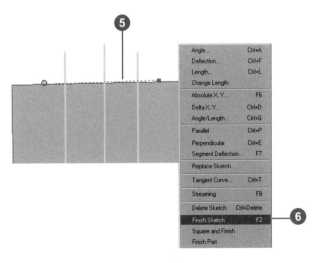

in which the vertices were added. If you looked to your right, you would be looking at the right side of the sketch.

The lines are trimmed on the right side of the line you constructed.

The lines are trimmed where the sketch was drawn.

Extending a line

The Extend task is the opposite of the Trim task, extending selected lines to a line you construct. Consider the roads and coastline example shown in the Trim task. If your database has some roads that should end at the coastline, but instead stop short, you could use the Extend task. By drawing a sketch on top of the coastline, you can extend the roads to the sketch you drew.

Tip

Other ways to construct a sketch

You can also use the Distance–Distance tool, the Arc tool, or the Intersection tool to create a sketch. For more information, see Chapter 4, 'Creating new features'.

1. Click the Current Task dropdown arrow and click Extend/Trim Features.

2. Click the Edit tool.

3. Click the line or lines you want to extend.

4. Click the tool palette dropdown arrow and click the Sketch tool.

5. Construct a line to which you want to extend the selected line or lines.

6. Right-click anywhere on the map and click Finish Sketch. ▶

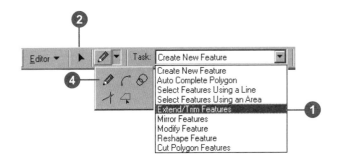

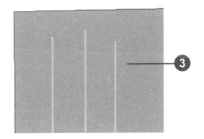

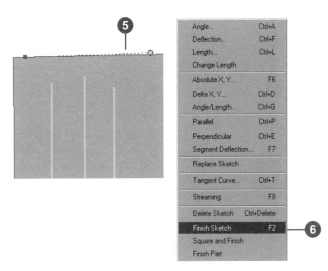

The lines are extended to the line you constructed.

The lines are extended to where the sketch was drawn.

Flipping a line

When you modify a line by trimming or extending it, the line is automatically trimmed or extended from its last vertex.

However, if you prefer to trim or extend a line from the first vertex instead of the last, you can use the Flip command. The Flip command reverses the direction of a line so that the last vertex of the sketch becomes the first.

1. Click the Current Task dropdown arrow and click Modify Feature.

2. Click the Edit tool.

3. Click the line whose direction you want to change.

4. Right-click over any part of the sketch and click Flip.

   The sketch becomes inverted (the first vertex becomes the last, marked in red).

5. When finished modifying the line, right-click over any part of the sketch and click Finish Sketch.

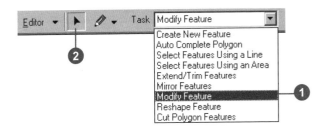

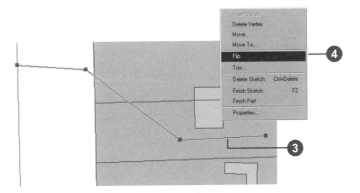

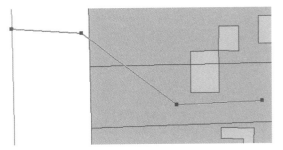

The first and last vertices of the line are reversed.

Placing points along a line

The Divide command creates points at a given interval along a line. For instance, you could use Divide to place utility poles along a primary.

You can create a specific number of points that are evenly spaced, or you can create points at a distance interval you choose.

1. Click the Edit tool.

2. Click the line you want to divide.

3. Click the Target layer dropdown arrow and click the point layer containing the type of points you want to place along the line.

4. Click Editor and click Divide. ▶

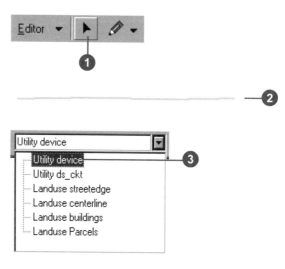

5. Click the first option and type a number to place a specific number of points evenly along the line.

Or click the second option and type a number to place the points at a specific interval in map units.

6. Click OK.

The line is divided by points placed along the line as specified.

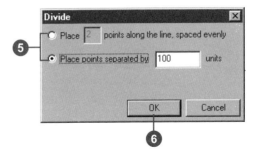

The line is divided by points.

Reshaping a line or polygon

The Reshape Feature task lets you reshape a line or polygon by constructing a sketch over the feature. The feature takes the shape of the sketch from the first place the sketch intersects the feature to the last.

When you reshape a polygon, if both endpoints of the sketch are within the polygon, the shape is added to the feature. ▶

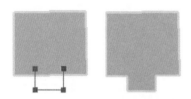

Tip

Other ways to construct a sketch

You can also use the Distance–Distance tool, the Arc tool, or the Intersection tool to create a sketch. For more information, see Chapter 4, 'Creating new features'.

1. Click the Current Task dropdown arrow and click Reshape Feature.

2. Click the Edit tool.

3. Click the feature you want to reshape.

4. Click the tool palette dropdown arrow and click the Sketch tool.

5. Create a line according to the way you want the feature reshaped.

6. Right-click anywhere on the map and click Finish Sketch. ▶

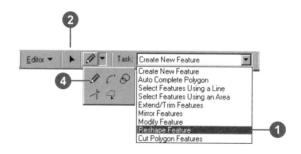

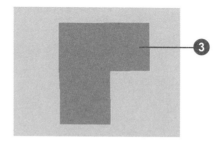

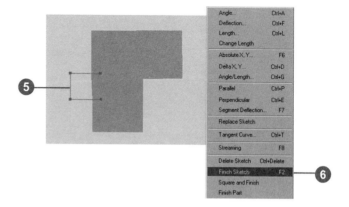

If the endpoints are outside the polygon, the feature is cut away.

When you reshape a line, both endpoints of the sketch must be on the same side of the line. The line takes the shape of the sketch you draw.

The feature is reshaped.

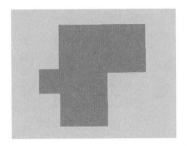

The feature is reshaped according to the sketch you constructed.

Adding and deleting sketch vertices

You can easily add *vertices* to or delete vertices from a sketch using the Insert Vertex and Delete Vertex commands on the Sketch context menu. By adding or deleting vertices, you can reshape a feature when you obtain new or better geographic data.

Suppose you have an existing layer with curb lines and receive an aerial photo that shows that the lines in the layer are incorrectly shaped. Using the aerial ▶

Tip

Adding vertices from the last vertex

You can add vertices to a feature beginning from the last vertex of the sketch. Click the Edit tool and double-click the feature to see its sketch. Then, click the Sketch tool to begin digitizing vertices.

Adding a vertex to a sketch

1. Click the Current Task dropdown arrow and click Modify Feature.

2. Click the Edit tool and click the line or polygon to which you want to add a vertex.

3. Move the pointer to where you want the vertex inserted and right-click.

4. Click Insert Vertex.

   A vertex is added to the sketch.

5. When finished modifying the line, right-click over any part of the sketch and click Finish Sketch.

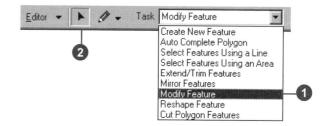

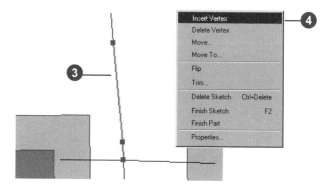

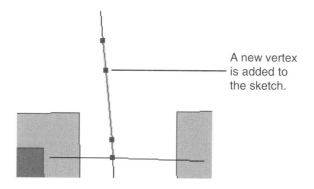

A new vertex is added to the sketch.

photo as a backdrop, you can add vertices to the curb lines as needed and then reshape the feature to match the photo by moving the vertices to new locations. You can also reshape the curb line features by deleting existing vertices from their sketches.

See Also

To learn how to move a vertex, see 'Moving a vertex in a sketch' in this chapter.

Deleting a vertex from a sketch

1. Click the Current Task dropdown arrow and click Modify Feature.

2. Click the Edit tool.

3. Click the line or polygon from which you want to delete a vertex.

4. Position the pointer over the vertex you want to delete until the pointer changes.

5. Right-click and click Delete Vertex.

   The vertex is deleted from the sketch.

6. Right-click over any part of the sketch and click Finish Sketch.

   The feature is reshaped.

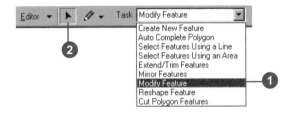

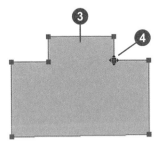

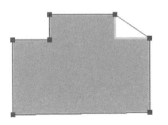

The vertex is deleted, and the feature is reshaped.

Moving a vertex in a sketch

Moving a vertex in a sketch offers another way to modify or reshape a feature.

ArcMap lets you move a vertex in several ways: by dragging it, by specifying new x,y coordinates, or by moving it relative to its current location.

You might choose to drag a vertex to a new location when you want to reshape a feature according to additional data you receive. For instance, you can drag a vertex to reshape a road feature in an existing layer in order to match it to the feature in a more accurate aerial photo.

You might move a vertex by specifying new x,y locations when you obtain additional data that provides the exact coordinate location at which the vertex should be. For example, suppose a parcel is resurveyed and a new GPS point is obtained for the parcel corner. You can move the corner of the parcel to match the location found by the GPS by specifying the equivalent location in x,y coordinates.

The Sketch context menu also provides a way to move a ▶

Dragging a vertex

1. Click the Current Task dropdown arrow and click Modify Feature.

2. Click the Edit tool and click the line or polygon whose vertex you want to move.

3. Position the pointer over the vertex you want to move until the pointer changes.

4. Click and drag the vertex to the desired location.

5. Right-click over any part of the sketch and click Finish Sketch.

   The feature is reshaped.

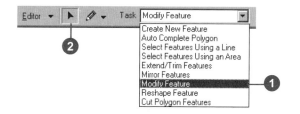

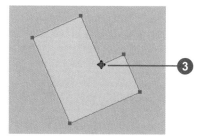

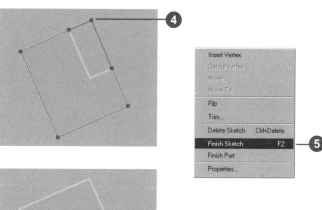

The vertex is moved, and the feature is reshaped.

vertex relative to its current location. Suppose an electrical pole must be moved 15 feet east and 5 feet north of its current location due to a road widening. Before moving the pole, you must reshape its electrical line so that the pole can connect to the line in the new location; you can do this by moving the vertex of the electrical line on which the pole sits using relative (delta) x,y coordinates.

The original location of the vertex as the origin (0,0) is used, and the vertex is moved to the new location using the map unit coordinates you specify (15,5 in this example). After the vertex is moved and the electrical line is reshaped, you can snap the pole feature to the vertex in its new location.

Maintaining a feature's shape when moving a vertex

You can also move a vertex without changing the shape of the feature. For more information, see 'Stretching a feature's geometry proportionately' in this chapter.

Moving a vertex by specifying x,y coordinates

1. Click the Current Task dropdown arrow and click Modify Feature.

2. Click the Edit tool and click the line or polygon whose vertex you want to move.

3. Position the pointer over the vertex you want to move until the pointer changes.

4. Right-click and click Move To.

5. Type the x,y coordinates where you want to move the vertex.

   The vertex is moved. ▶

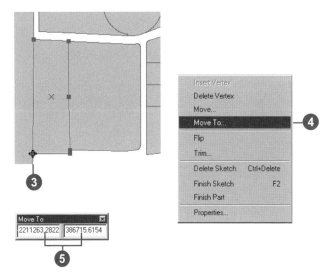

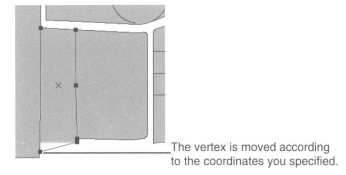

The vertex is moved according to the coordinates you specified.

Undoing a vertex move

If you move a vertex and don't want it to stay in the new location, click the Undo button on the ArcMap Standard toolbar. The vertex returns to its last position. Click the Redo button if you want to move the vertex back to the new location.

6. Right-click over any part of the sketch and click Finish Sketch.

The feature is reshaped.

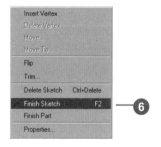

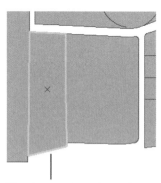

The feature is reshaped.

Moving a vertex relative to its current location

1. Click the Current Task dropdown arrow and click Modify Feature.

2. Click the Edit tool and click the line or polygon whose vertex you want to move.

3. Position the pointer over the vertex you want to move until the pointer changes.

4. Right-click and click Move.

5. Type the delta x,y coordinates where you want to move the vertex. ▶

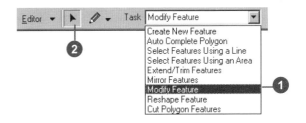

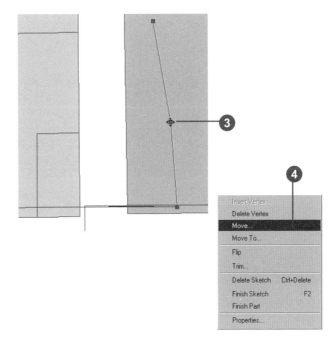

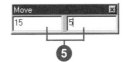

The vertex is moved.

6. Right-click over any part of the sketch and click Finish Sketch.

The feature is reshaped.

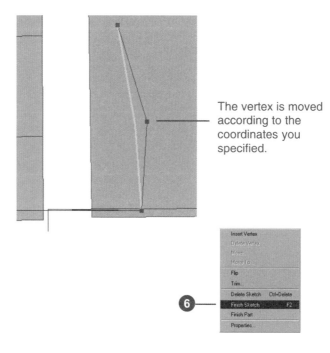

The vertex is moved according to the coordinates you specified.

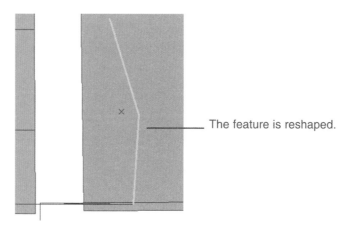

The feature is reshaped.

Changing the properties of a sketch

When creating a new feature or modifying an existing one, you can easily change the properties of the sketch shape using the Sketch Properties dialog box.

Using the Sketch Properties dialog, you can remove parts from a multipart feature, insert and delete vertices, and alter the m- and z-values of vertices.

Suppose you are editing a layer that contains river features whose shapes contain too many vertices. You could use the Sketch Properties dialog box to select unwanted vertices and delete them.

Tip

How do I know which vertices I have selected?

As you select vertices in the dialog box, they turn orange on the map.

Deleting multiple vertices from a feature

1. Click the Edit tool and select the feature whose shape you want to modify.

2. Click the Current Task dropdown arrow and click Modify Feature to place the shape of the feature in the edit sketch.

3. Right-click the sketch and click Properties.

4. Select the vertices that you want to remove by holding down the Shift key and clicking vertices from the table. Use the Shift and Ctrl keys to select more than one vertex.

5. Click the Delete key or right-click over the selected vertices and click Delete.

   The vertices form the sketch.

6. Click Finish Sketch.

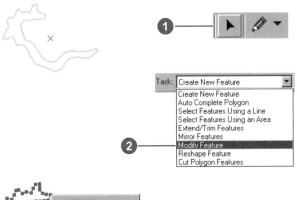

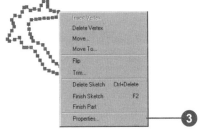

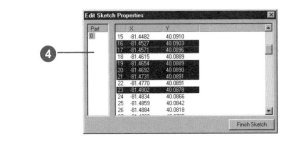

Tip

Tip

Modifying the x,y coordinates

If you don't want the added point to be exactly at the midpoint between two vertices, click the x or y column and type in a new coordinate for the point.

Tip

Insert vertices after a selected vertex

You can insert vertices either before or after the vertex that you right-click on top of.

Inserting a vertex at the midpoint of a segment

1. Right-click over a segment of the edit sketch and click Properties.

2. Select the vertex before which you wish to insert a new vertex.

3. Right-click the selected vertex and click Insert Before.

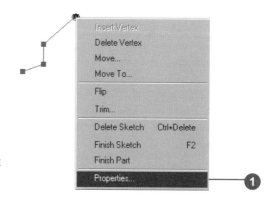

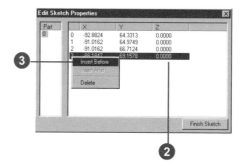

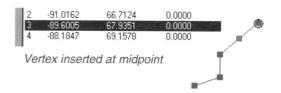

Vertex inserted at midpoint

How do I know which parts I have selected?

When you select a part from the Sketch Properties dialog box, the segments for that part will appear thicker.

Removing a part from a multipart feature

1. Click the Edit tool and select the feature you want to remove a part from.

2. Click the Current Task dropdown arrow and click Modify Feature to place the multipart shape in the edit sketch.

3. Right-click the sketch and click Properties.

4. Right-click the part that you want to remove and press the delete key or right-click and click Delete.

5. Click Finish Sketch.

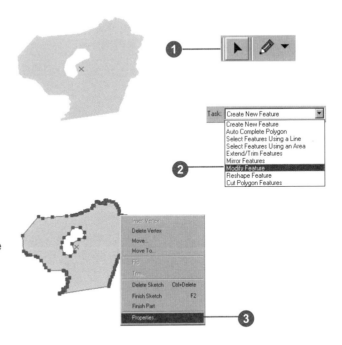

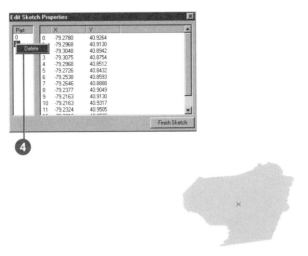

Using the Current Z control

When you add points to the edit sketch, you can control the z-value for each vertex using the Current Z tool.

To use the Current Z tool, you must first add it to a toolbar from the Commands tab of the Customize dialog box. The Current Z tool is listed in the Editor category.

Editing z- and m-values of a feature

1. Click the Edit tool and select the feature whose z- or m-values you wish to edit.

2. Click the Current Task dropdown arrow and click Modify Feature.

3. Right-click on top of the sketch and click Properties.

4. Select the vertex you wish to modify.

5. Click the z or m field in the table and type a new value.

6. Click Finish Sketch.

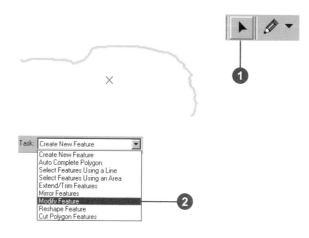

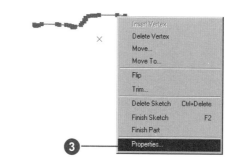

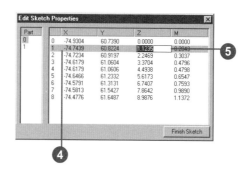

Scaling features

You can scale a feature—make the entire feature larger or smaller—using the Scale tool. The feature is scaled based on the location of the selection anchor—the small "x" located in the center of selected features.

You might use the Scale tool when working with data from a new source in which the scale is slightly inaccurate—for example, subdivision parcels from a surveyor. You can use the Scale tool to scale parcels so that they fit together properly.

To use the Scale tool, you must first add it to a toolbar from the Commands tab of the Customize dialog box. The Scale tool is available from the Editor category. For more information on adding a tool to a toolbar, see *Exploring ArcObjects*.

Tip

Moving the selection anchor

To move the selection anchor of a feature you want to scale, hold the scaling pointer over the anchor until the icon changes. Then, click and drag the anchor to a new location.

1. Click the Edit tool.
2. Click the feature you want to scale.
3. Click the Scale tool.
4. Move the selection anchor if necessary.
5. Click and drag the pointer over the feature to scale it as desired. ▶

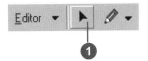

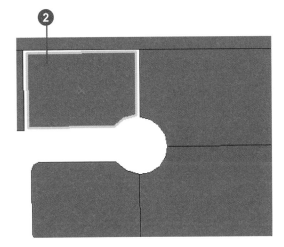

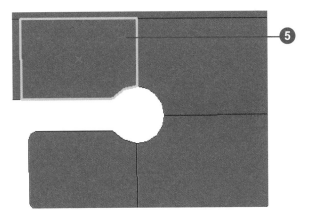

Tip

Scaling more than one feature

You can scale more than one feature at the same time. Simply select all the desired features and move the selection anchor to the desired location before using the Scale tool.

Tip

Undoing scaling

To return a feature to its original size after scaling it, click the Undo button on the ArcMap Standard toolbar.

Tip

Scale factor

You can scale features using a scale factor instead of dragging the mouse. Click the F key to set the scale factor.

6. Release the mouse button when you're finished scaling the feature.

The feature is scaled.

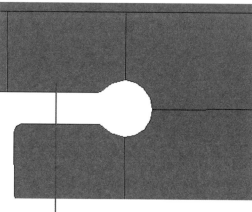

The feature is scaled.

Clipping features

You can easily clip features that touch or are within a buffered distance of selected features.

Suppose that you want to model the effect of a proposed road-widening project on the lots of a subdivision block. You can do this using the Clip command. Select the road centerline where the proposed widening is to occur and then click Clip from the Editor menu. Type the length measurement of the widening and click the option to Discard the area that intersects to clip the subdivision lots.

When using the Discard the area that intersects option, the Clip command will buffer the selected road feature and then clip all portions of editable features that are within the buffered region. Using the "Preserve the area that intersects" option, all features that touch the buffered feature will be deleted.

1. Select the feature you want to use to clip features.

2. Click Editor and click Clip.

3. Type a buffer value. You can leave the value as 0 if you are using a polygon feature to clip with.

4. Click the type of clip operation you wish to use.

5. Click OK to clip the feature.

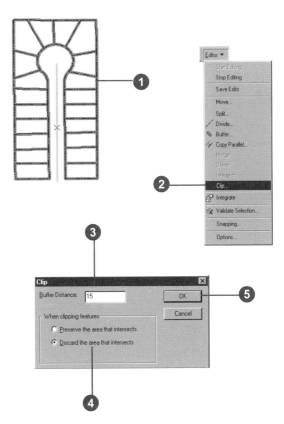

Stretching geometry proportionately

Sometimes you want to stretch a feature without changing its geometry (shape). Suppose you want to change the position of a feature in relation to other features by moving a vertex. For example, perhaps the data you have for an electric transmission system is not as accurate as you would like. However, you have other layers containing accurate surveyed points that coincide with some of the transmission towers, power generating plants, and substations. By moving the vertices of the transmission lines, you can adjust the positions of the lines to the known surveyed positions of the features in the more accurate layer. You can change the positions of these vertices without changing the general shape of the transmission lines by stretching the features proportionately.

When you stretch a feature proportionately, the proportions of the feature's segments are maintained, thereby maintaining the general shape of the feature. This is different from moving a vertex to reshape a feature.

The graphics below show the difference between moving a vertex to reshape a feature and moving a vertex while maintaining the shape of the feature. The three graphics on the top show how a feature is modified when its upper-right vertex is moved with proportionate stretching turned on. The three graphics on the bottom show how the same feature is reshaped when its upper-right vertex is moved with proportionate stretching turned off.

Proportionate stretching on

Proportionate stretching off

 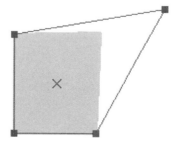 

Stretching a feature's geometry proportionately

Within the Editing Options menu, you can choose to stretch the geometry of features proportionately when moving vertices. When you drag a vertex to a new location with this option turned on, the proportions of the feature's segments are maintained, thereby maintaining the general shape of the feature.

You might want to stretch features proportionately when merging data from different data sources—for example, utility lines from one source and subdivision parcels from another.

Suppose the data for the subdivision parcels is very accurate, but the data for the utility lines is not as accurate. While the shapes of the utility lines are generally correct, you want to change the position of one line relative to the parcels by moving a vertex. By stretching the utility line feature proportionately, you can make it fit accurately with the parcels without losing the general shape of the line. ▶

1. Click the Current Task dropdown arrow and click Modify Feature.

2. Click the Edit tool and click the feature you want to stretch.

3. Click Editor and click Options.

4. Click the General tab.

5. Check the check box to stretch the feature proportionately.

   Uncheck the check box if you want to reshape the feature without maintaining proportionate geometry.

6. Click OK. ▶

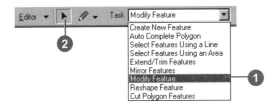

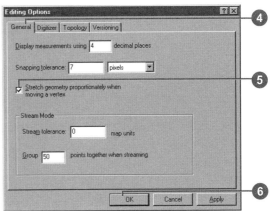

You can turn on proportionate stretching by checking a box on the General tab of the Editing Options dialog box. Uncheck the box if you simply want to reshape a feature without maintaining proportionate geometry.

See Also

To see how stretching a feature proportionately looks in comparison to stretching a feature to reshape it, see 'Stretching geometry proportionately' in this chapter.

7. Position the pointer over the vertex you want to move until the pointer changes.

8. Drag the vertex to the desired location.

9. Right-click over any part of the sketch and click Finish Sketch.

   The feature is stretched proportionately.

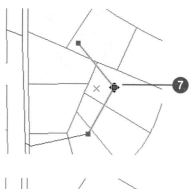

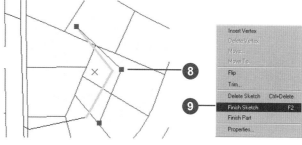

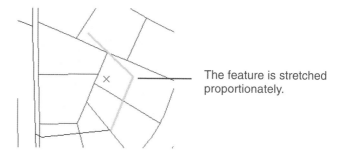

The feature is stretched proportionately.

Editing topological features

8

Most vector datasets—whether they are shapefiles, coverages, or feature datasets—have features that share boundaries or corners. Editing a boundary or vertex shared by two or more features updates the shape of each of those features. This is called a *topological association*.

A topological association means that some parts of the features' shapes share the same location. In addition, different feature classes in a feature dataset often share geometry between them. For example, moving a slope boundary in one feature class would also update a forest stand in another feature class.

The topology tools in ArcMap let you maintain topological associations between features and between feature classes when you edit shapefiles, coverages, or feature datasets. In this chapter, you will learn how to use the topology tools as well as how to create topologically integrated data.

Topological associations in a geometric network are not discussed here; they are covered in *Building a Geodatabase*. For detailed information about topology in general, see *Modeling Our World*. Information specific to editing coverage topology is covered in Appendix A, 'Editing coverages in ArcMap'.

Integrating topological data

Before you edit data that has topological associations, the data must be topologically integrated so that all features with parts that should be shared are shared.

Automatic integration

When you begin to edit topological data with the topology tools in ArcMap, the tools will automatically integrate the area you are editing if that area has not already been integrated. This involves checking all feature classes in your feature dataset and making any boundaries or vertices within a certain distance range identical, or *coincident*. This distance range is called the cluster tolerance. The *cluster tolerance* determines the range within which features are made coincident. When you edit data with the topology tools, ArcMap always integrates the data at the smallest possible cluster tolerance value.

Manually integrating data

You can also manually integrate shapefiles and geodatabase feature datasets using the Integrate command. You might manually integrate data in this way if you want to specify a cluster tolerance that is different from the one used to automatically integrate the data.

For example, suppose you specify a cluster tolerance of 5 map units. Suppose your data had a parcel boundary that should be shared with the adjacent parcel boundary but was 4 map units away. After running Integrate, the boundaries of the two parcels would be made coincident because they were within the cluster tolerance of 5 map units.

To minimize error, the cluster tolerance you choose should be as small as possible, depending on the precision level of your data. For example, if your data is accurate within 10 meters, you would want to set your cluster tolerance no larger than 10 meters and smaller if possible. The default cluster tolerance is 0, representing the minimum possible tolerance value. This default value is applied if you don't specify a cluster tolerance.

Integrating data

Before you edit data with topological associations, it must be topologically integrated.

As discussed in the previous section, ArcMap automatically integrates topological data that is not already integrated when you begin to edit it with the topology tools.

However, if you wish to manually integrate shapefiles or feature datasets in a geodatabase, you can use the Integrate ▶

Tip

When should I manually integrate my data?

Use the Integrate command when you want to integrate your shapefile or geodatabase feature dataset using a cluster tolerance that is different from the minimum tolerance ArcMap uses when it automatically integrates your data.

See Also

For more information on choosing a cluster tolerance, see 'Integrating topological data' in this chapter.

See Also

For a description of the different types of data you can edit in ArcMap, see Chapter 3, 'Editing basics'.

1. Add the shapefile or feature dataset you want to integrate to your map.

2. Click the Editor menu and click Start Editing.

3. Click Editor and click Options.

4. Click the Topology tab.

5. Uncheck the check box only if you want to integrate the entire dataset.

   Leave it checked if you want to integrate features in the current map extent.

6. Type the distance range (in map units) within which you want features to be coincident.

7. Click OK. ▶

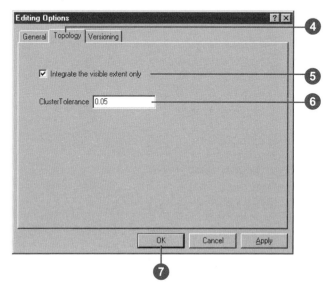

command on the Editor menu. This involves setting a cluster tolerance and specifying whether you want to integrate the entire dataset or only the features in the current map extent.

The Integrate the visible extent only check box—on the Topology tab of the Editing Options dialog box—is checked by default to integrate only the features in the current map extent. Uncheck the check box only if you want to integrate the entire dataset.

When you integrate a feature dataset, you are integrating all of the feature classes it contains, even the ones that are not visible on the map. You should therefore make sure that you know the contents of your dataset before you integrate it.

You cannot manually integrate coverages. When you use the topology tools to edit a coverage that is not already integrated, the coverage is integrated automatically using the smallest possible cluster tolerance.

Tip

"Undoing" an Integrate
You can undo an Integrate by clicking the Undo button on the ArcMap Standard toolbar.

8. Click Editor and click Integrate.

   The data is integrated.

Moving a shared vertex or boundary

Use the Shared Edit tool to move parts of features that are shared. This tool also selects parts of features on which you want to use other topology tools. The Shared Edit tool is designed to select vertices or boundaries that are shared between two or more features.

When you select with the Shared Edit tool, only the topmost visible vertex or boundary on the map is highlighted. However, every vertex or boundary in the dataset underneath where you clicked on the map is actually selected. This ensures that when you move part of a feature that is shared, any coincident vertices or boundaries underneath will move appropriately, as well as any vertices or boundaries connected to those parts. This is true even for feature classes that aren't visible on the map.

Moving a shared vertex

1. Click the Shared Edit tool.

2. Click the shared vertex you want to move.

   The selected vertex is highlighted.

3. Drag the vertex to the desired location.

   The location of the vertex is updated, along with all coincident vertices or boundaries in the dataset, as well as any connected vertices or boundaries.

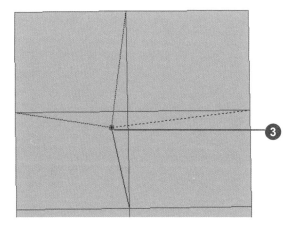

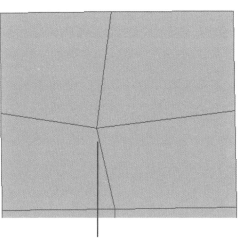

The vertex, all coincident vertices and boundaries in the dataset, and any connected vertices and boundaries are moved.

"Undoing a shared edit"

You can undo a shared edit by clicking the Undo button on the Standard toolbar. If the edit happened in an area that was automatically integrated, you need to click Undo twice to remove the automatic integration.

Moving a shared boundary

1. Click the Shared Edit tool.

2. Click the shared boundary you want to move.

   The selected boundary is highlighted.

3. Drag the boundary to the desired location.

   The location of the boundary is updated, along with all coincident vertices or boundaries in the dataset, as well as any connected vertices or boundaries.

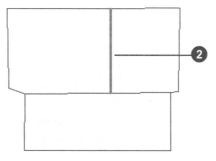

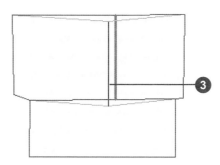

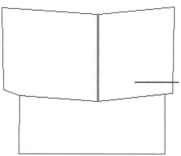

The boundary, all coincident vertices and boundaries in the dataset, and any connected vertices and boundaries are moved.

Reshaping a shared boundary

You can use the Shared Edit tool to select a shared boundary you want to reshape. The Reshape Feature task lets the boundary take the shape of a sketch you draw from the first intersection to the last. The sketch must intersect the boundary at least twice.

When you reshape a shared boundary, all coincident vertices and boundaries, as well as any connecting vertices or boundaries, are reshaped as well.

See Also

For more information on reshaping a feature using a sketch, see 'Reshaping a line or polygon' in Chapter 7.

1. Click the Current Task dropdown arrow and click Reshape Feature.

2. Click the Shared Edit tool.

3. Click the shared boundary you want to reshape. ▶

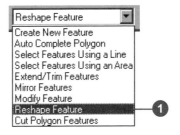

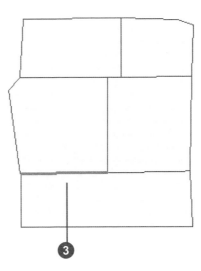

Other ways to construct a sketch

You can also use the Distance–Distance tool, the Arc tool, or the Intersection tool to create a sketch. For more information, see Chapter 4, 'Creating new features'.

4. Click the tool palette dropdown arrow and click the Sketch tool.

5. Create a line according to the way you want the feature reshaped.

6. Double-click the last vertex to finish the sketch.

    The boundary and all coincident vertices and boundaries, as well as any connecting vertices or boundaries, are reshaped.

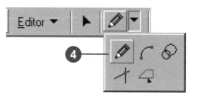

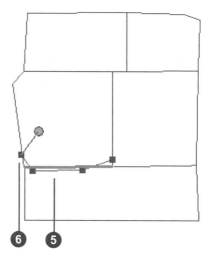

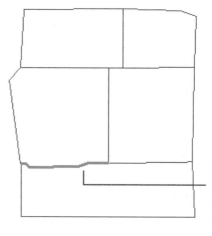

The boundary, all coincident vertices and boundaries, and any connecting vertices and boundaries are reshaped.

Modifying a shared boundary

You can use the Shared Edit tool to select a shared boundary that you want to modify. Then, use the Modify Feature task to see the sketch of the boundary.

You can modify the sketch boundary as you would any feature sketch—by inserting, deleting, and moving vertices using the Sketch context menu.

When you modify the boundary selected with the Shared Edit tool, all coincident vertices and boundaries as well as any connecting vertices or boundaries are modified accordingly.

Tip

Shortcut for modifying shared boundaries

Instead of using the Modify Feature task to change a shared boundary, you can click the Shared Edit tool and double-click the shared boundary you want to modify.

See Also

For more information on modifying features, see Chapter 7, 'Editing existing features'.

1. Click the Current Task dropdown arrow and click Modify Feature.

2. Click the Shared Edit tool.

3. Click the shared boundary you want to modify.

   The sketch of the boundary appears. ▶

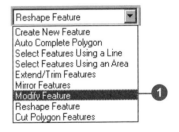

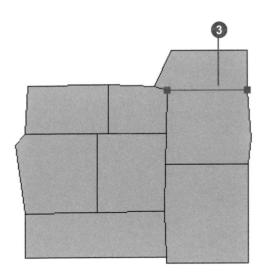

4. Modify the sketch as desired.

5. Right-click anywhere over the boundary and click Finish Sketch.

   The boundary and all coincident vertices and boundaries, as well as any connecting vertices or boundaries, are modified accordingly.

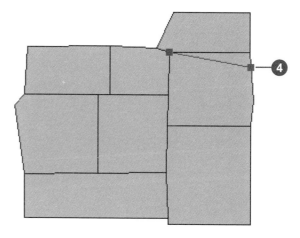

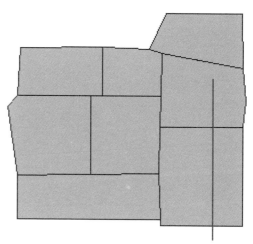

The boundary, all coincident vertices and boundaries, and any connecting vertices and boundaries are modified.

Creating a new polygon with shared parts

Use the Auto Complete Polygon task to create a new polygon if it shares part of its boundary with one or more existing polygons. This way, it's not necessary to re-create the portion of the polygon that's already represented.

When you create a new polygon in this way, all boundaries and vertices of the polygon are automatically shared with the existing polygons.

Tip

Using the snapping environment to create a sketch

You can use the snapping environment to help you snap to existing polygon boundaries when creating a sketch. For more information on snapping, see Chapter 4, 'Creating new features'.

1. Click the Current Task dropdown arrow and click Auto Complete Polygon.

2. Click the Target layer dropdown arrow and click a polygon layer.

3. Click the Sketch tool.

4. Create a sketch that starts and stops at any of the existing polygon boundaries to enclose the new polygon you are creating.

   You can overshoot the existing boundaries when you create the line; they will be trimmed automatically.

5. Double-click to finish the sketch. ▶

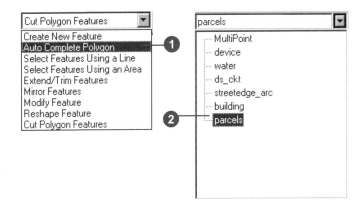

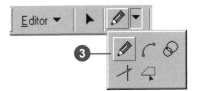

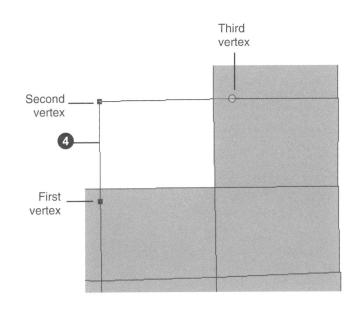

Any vertices or boundaries necessary to complete the polygon are created and automatically shared. Any overshoots from the sketch you drew are removed.

All the vertices and boundaries of the polygon are created and automatically shared. Overshoots from the sketch are removed.

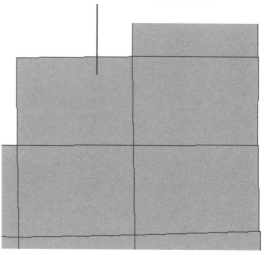

Editing attributes

9

IN THIS CHAPTER

- **Viewing attributes**

- **Adding and modifying attributes**

- **Copying and pasting attributes**

GIS database editing not only involves creating and editing features; it also involves assigning and updating the *attributes*, or characteristics, of features.

ArcMap makes it easy to view and update the attributes of features in your database. You can edit feature attributes in two ways: using the Attributes dialog box or using a feature layer's attribute table. This chapter focuses on editing attributes using the Attributes dialog box. With the Attributes dialog box, you can view the attributes of selected features on your map; add, delete, or modify an attribute for a single feature or multiple features at the same time; and copy and paste individual attributes or all the attributes of a feature.

You can perform similar functions using a feature layer's attribute table. However, with tables you can also do computations—such as adding and sorting records—with attribute values. To learn how to edit attributes in an attribute table—including performing computations with attribute values—see *Using ArcMap*.

Viewing attributes

The Attributes dialog box lets you view the attributes of features you've selected in your map. The left side of the dialog box lists the features you've selected. Features are listed by their primary display field and grouped by layer name. The number of features selected is displayed at the bottom of the dialog box.

The right side of the Attributes dialog box is called the property inspector. The property inspector contains two columns: the attribute properties of the layer you're viewing, such as Type or Owner, and the values of those attribute properties.

Tip

Finding the feature on the map

You can find a selected feature on the map by either highlighting or zooming to it. To highlight the feature, click the primary field and the feature will flash on the map. Right-click the field and click Zoom To in the context menu to get a close-up view of the feature. Click the Back button on the Tools toolbar to return to the previous map extent.

1. Click the Edit tool.
2. Select the features whose attributes you want to view.
3. Click the Attributes button.
4. Click the layer name that contains the features whose attributes you want to view.

   The layer's attribute properties appear on the right side of the dialog box. ▶

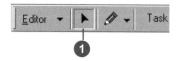

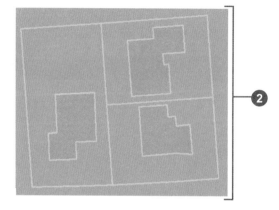

The layer's attribute properties appear.

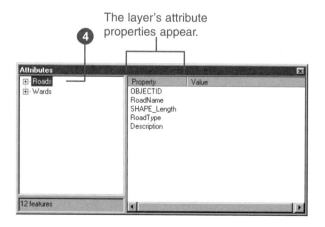

Tip

Unselect a feature

You can remove features from the selection without having to click on the map. To remove a feature from the selection, right-click on the feature and click Unselect from the context menu.

Tip

Delete a selected feature

If you want to delete a feature without losing your selection, simply right-click on the feature and click Delete from the context menu.

Tip

Viewing attributes in the Identify Results window

To view the attributes of a feature quickly, click the Identify Features button on the Tools toolbar, then click the feature whose attributes you want to view. View the feature's attributes in the Identify Results window.

Tip

Changing the primary display field

You can change the primary display field for a layer on the Fields tab of the layer's Properties dialog box. To open the dialog box, right-click the layer name in the table of contents.

5. Double-click a layer name to see the primary display fields, representing the selected features in the layer.

   Double-click again to hide the primary display fields.

6. Click a primary display field to see the corresponding feature's attribute values.

   The corresponding feature flashes on the map.

7. Click the Close button to close the dialog box.

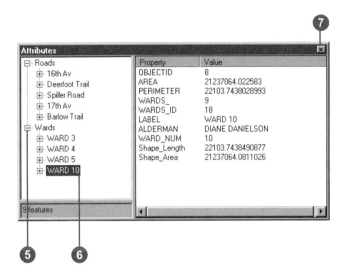

Adding and modifying attributes

The easiest way to make changes to the attributes of a selected feature is by using the Attributes dialog box.

You can add or modify attributes of selected features as needed. For example, you might want to update the attribute values—such as its name and maintenance information—for a park feature you created.

To add or modify an attribute value for a single feature, click the primary display field for the feature on the left side of the dialog box and make your changes in the Value column on the right. ►

Tip

Saving your edits
Click the Editor menu and click Save Edits.

Tip

Attribute domains
You can use attribute domains to create a list of valid values for a feature in a geodatabase. You can also use the Validate command to ensure attribute quality. For more information, see Building a Geodatabase.

Adding an attribute value to a single feature

1. Click the primary display field of the feature to which you want to add an attribute value.

2. Click in the Value column where you want to add the attribute value.

3. Type the attribute value and press Enter.

   The attribute value is added to the feature.

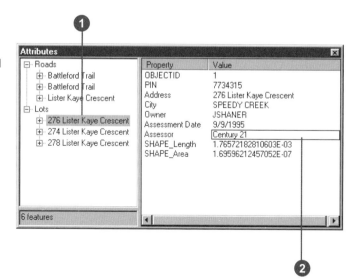

Adding an attribute value to all selected features in a layer

1. Click the layer to which you want to add an attribute value.

2. Click in the Value column where you want to add the attribute value.

3. Type the attribute value and press Enter.

   The attribute value is added to all selected features in the layer.

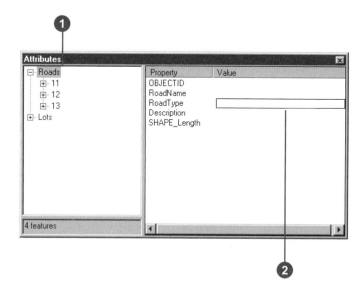

You can also add or modify an attribute value for all selected features in a layer at the same time. Simply click the layer name on the left and make your changes in the Value column on the right.

Tip

Deleting attributes

To delete an attribute value, right-click over the value and click Delete. You can also press the Delete key to delete an attribute value.

Tip

Undoing your edits

To undo any edit to feature attributes, click the Undo button on the ArcMap Standard toolbar.

Tip

Performing calculations

When editing attributes, you might need to perform calculations using the field calculator in the feature layer's attribute table dialog box. For more information, see Using ArcMap.

Tip

Adding attribute properties

You can add an attribute property for a feature—for example, Owner or Type—by working with its attribute table in ArcCatalog. For more information, see Using ArcCatalog.

Modifying an attribute value for a feature

1. Click the primary display field of the feature for which you want to modify an attribute value.

2. Click the value you want to modify.

3. Type a new attribute value and press Enter.

   The attribute is modified for the feature.

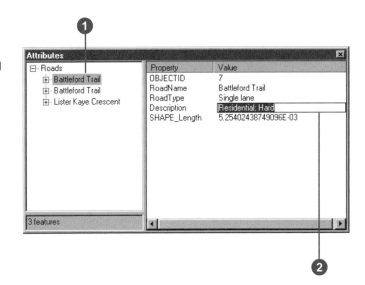

Modifying an attribute value for all selected features in a layer

1. Click the layer for which you want to modify an attribute value.

2. Click in the Value column next to the attribute property you want to modify for all selected features in the layer.

3. Type a new attribute value and press Enter.

   The attribute is modified for all selected features in the layer.

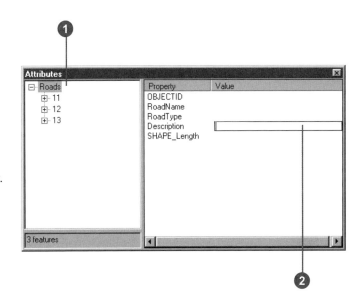

Copying and pasting attributes

Copying and pasting is an easy way to edit the attributes of features on your map. You can copy individual attribute values or all the attribute values of a feature. Attribute values can be pasted to a single feature or to all selected features in a layer.

Tip

Copying and pasting individual attribute values to an entire layer

To copy an attribute value to a layer, click the value you want to copy, right-click, and click Copy. Then, click the layer name and right-click in the Value column next to the appropriate property. Click Paste and the attribute value is copied to every selected feature in the layer.

Tip

Cutting and pasting attributes

Cutting and pasting attributes is similar to copying and pasting them. Right-click and click cut from the context menu to remove the attribute value from its current location in the Attributes dialog box, then click Paste to paste it elsewhere.

Copying and pasting individual attribute values from feature to feature

1. Click the attribute value you want to copy.

2. Right-click the value you want to copy and click Copy.

3. Click the primary display field of the feature to which you want to paste the value.

4. Click where you want to paste the value.

5. Right-click where you want to paste the value and click Paste.

   The attribute value is pasted to the feature.

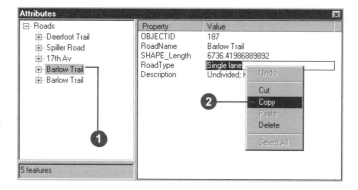

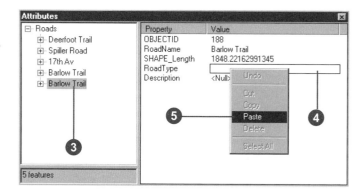

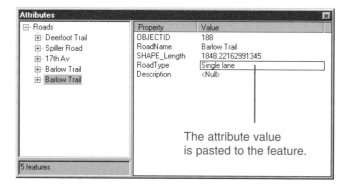

The attribute value is pasted to the feature.

Copying and pasting all attribute values from feature to feature

1. Right-click the primary display field of the feature whose attribute values you want to copy and click Copy.

2. Right-click the primary display field of the feature to which you want to paste the attribute values and click Paste.

   The attribute values are pasted to the feature.

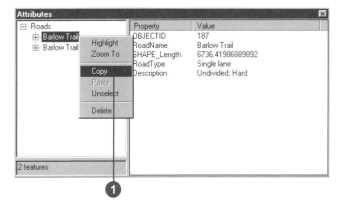

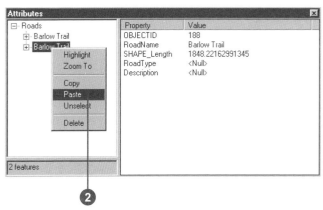

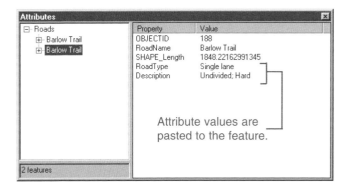

Attribute values are pasted to the feature.

Editing coverages in ArcMap

ArcMap lets you edit ArcInfo coverages, ESRI shapefiles, personal geodatabases, and multiuser geodatabases. It was designed to be easy to use and learn. As much as possible, ArcMap provides a normalized approach to editing all of these data sources. A key part of this design is that features from these data sources are presented to ArcMap as objects. When editing coverage data, users familiar with ARCEDIT™ should be aware that there are several basic differences in the approach of these systems.

You will find that the basic patterns of editing geographic data are present in ArcMap, and much of what you know about editing topology will help you learn to edit coverages quickly. Please read Chapter 8, 'Editing topological features', to learn more about topological editing.

This appendix describes the underlying coverage data model. It guides you through the process of creating coverage features and then outlines each editing task that is unique to coverage features.

Coverages and topology

Coverages use a collection of *feature classes* to represent geographic features; each feature class stores a set of points, lines (arcs), polygons, or annotation (text). To define the geographic features, more than one feature class is often required. For example, both line and polygon feature classes exist in a coverage representing land use areas. Polygon features also have label points, which are stored in a separate feature class.

Feature classes can be separated into three distinct types: primary, secondary, and composite.

Primary feature classes include points (label points), lines (arcs), polygons, and nodes. When node and polygon feature classes are present in a coverage, the coverage has topology and the features in those classes contain relationships, called *topological associations*, with other features: arcs form the perimeter of polygons, nodes are located at the endpoint of arcs, and label points mark the interiors of polygons.

These topological associations between features are stored directly in the feature class tables. For example, the topological association between a node feature and an arc feature is defined by the ARC#, FNODE#, and TNODE# items.

ARC#	FNODE#	TNODE#
3	4	5
4	5	6

This type of association is often referred to as "Arc–node topology".

Likewise, the topological association between an arc feature and a polygon feature is defined by the ARC#, LPOLY#, and RPOLY# items.

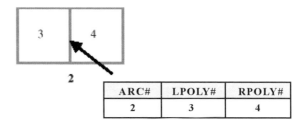

ARC#	LPOLY#	RPOLY#
2	3	4

This type of association is often referred to as "Polygon–arc topology".

In order to maintain topological associations between features while editing in ArcMap, you must use the topology tools. Using the Shared Edit tool, you can reshape the boundary of a polygon, and the shape of the arc will be updated automatically, thus maintaining the association between features. For more information about using topology tools, see Chapter 8, 'Editing topological features'.

Secondary feature class types include tics, links, annotation, and sections. Tics are used for map registration, and links are used for adjusting features. Sections are the building blocks of a measurement system that is stored on composite route features; annotation is text labels displayed on a map. Secondary feature classes cannot be edited in ArcMap.

Coverages also contain *composite* feature classes called subclasses. Routes are collections of arcs with an associated measurement system, while regions are collections of polygons, which can be adjacent, disjoint, or overlapping.

ArcMap and coverage feature class types

ArcMap identifies all coverage features as either simple features or topo features.

Simple features do not contain topological associations with other features. For example, well locations stored in a single-point feature class coverage are simple features. Coverages that contain simple features do not store topology. All editing tools can be used to create and update simple features.

Topo features store and maintain topological associations with other features in a coverage. For example, parcels stored in a polygon feature class maintain an association with line features stored in the arc feature class. You can attach attributes such as survey measurements or line type to the line features and model parcel ownership or tax value on the polygon features. Coverages that contain node and/or polygon features store topology.

ArcMap ensures that those topological associations are maintained by disabling editing tools that could invalidate the relationship. For example, when you select a polygon feature using the Edit tool, the Rotate command is disabled.

When you want to edit features that contain topological associations with other features, you need to select the boundaries and points that are shared between features. You can select and edit shared boundaries using the Shared Edit tool. For more information about the Shared Edit tool and editing topological features, see Chapter 8, 'Editing topological features'.

The table below identifies feature types based on the number and types of feature classes that are present in a coverage.

Coverages and feature types

Feature classes	Feature type
Points only, arcs only	Simple
Points and arcs	Simple
Arcs and nodes	Topo
Polygons only	Topo
Arcs and polygons	Topo
Arcs, polygons, and region subclasses	Topo
Arcs and route subclasses	Topo
Arcs, polygons, region, and route subclasses	Topo

The next topic guides you through the process of creating each type of coverage feature that can exist in a coverage. It explains how topological associations are created and maintained for each topo feature that you create.

The topic that follows describes editing tasks that are specific to topo features. For example, suppose you want to merge two polygons together. You need to select and delete the shared boundary between them using the Shared Edit tool—if you select both polygons using the Edit tool, the Merge command will be disabled. Editing tasks that are unique to coverage features are described in detail.

Creating new coverage features

The steps for creating a coverage feature may differ, depending on the type of coverage feature you want to create.

To create primitive features (points, arcs, and polygons), use the Create new feature task and digitize a shape using the sketch tools.

If the primitive feature is a topo feature, related features may be created to support the construction. For example, if you create a new polygon feature, new arc features and a label point feature will be created automatically.

To create composite features (routes and regions) you need to use the Union command.

Tip

Coverage tolerances

When creating new arc features, use snapping to ensure arc–node connectivity. Coverage editing tolerances (NodeSnap, ArcSnap, Weed, and Snapping) do not work in ArcMap. See 'Using the snapping environment' in Chapter 4.

Creating an arc feature

1. Click the Current Task dropdown arrow and click Create New Feature.

2. Click the Target layer dropdown arrow and click a coverage line layer.

3. Click the tool palette dropdown arrow and click the Sketch tool.

4. Click the map to create a new line sketch.

5. Double-click to finish the sketch and create a new line feature.

   If the arc coverage contains node features, nodes will be created at the endpoints of the line.

   Digitizing an arc that crosses the boundary of a polygon that also belongs to the same coverage will split the polygon in two.

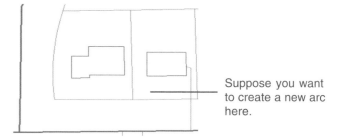

Suppose you want to create a new arc here.

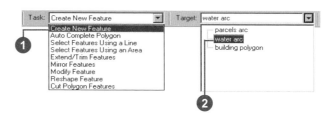

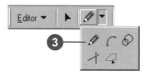

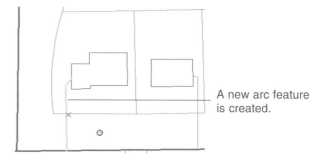

A new arc feature is created.

Creating a node

Coverage node features connect arc features together. When you create a new arc feature, node features are automatically placed at the endpoints of the arc feature. However, you can use the Create New Feature task to add nodes to your coverage. When you create a new node feature the node must touch or intersect an existing arc feature.

If you need to connect arc features together but do not have a node feature class in your coverage, you must create one using ArcCatalog prior to editing. You create node feature classes by generating arc–node topology. To learn more about creating coverage topology, see *Using ArcCatalog*.

Tip

Use snapping when creating node features

You can use snapping to ensure that nodes snap to existing arc features. For more information about snapping, see 'Using the snapping environment' in Chapter 4.

1. Click the Current Task dropdown arrow and click Create New Feature.

2. Click the Target layer dropdown arrow and click a coverage node layer.

3. Click the tool palette dropdown arrow and click the Sketch tool.

4. Click an existing arc to add a new node feature.

Suppose you want to create a new node here.

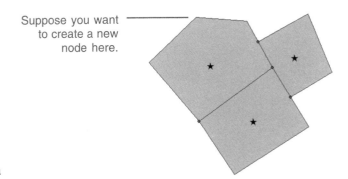

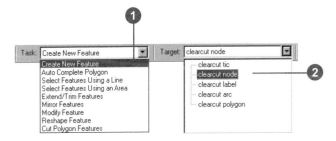

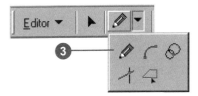

A new node feature is created, and the arcs are split.

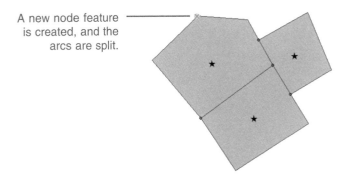

Creating point and label features

Simple point and label point features do not participate in the topology of a coverage. However, label points are used as placeholders for the attributes of polygons. You can create new points and label points using the Create New Feature task.

When you create a new coverage polygon, a label point is added automatically. You can add multiple label points inside of a polygon. Multiple label points can be used for advanced label placement. Each point will contain the same attribute information as the polygon feature and any other label points inside of that polygon.

1. Click the Current Task dropdown arrow and click Create New Feature.

2. Click the Target layer dropdown list and click a coverage point or label layer.

3. Click the tool palette dropdown arrow and click the Sketch tool.

4. Click to add the point feature.

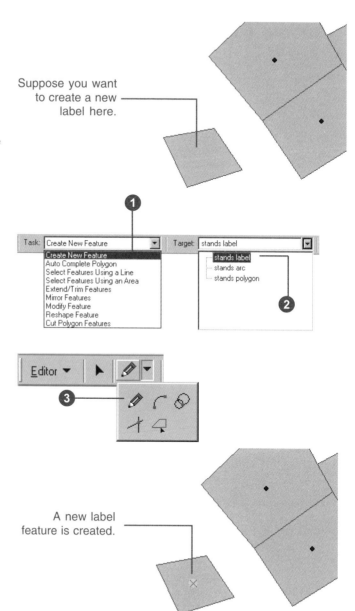

Suppose you want to create a new label here.

A new label feature is created.

Creating polygon features

You can create new coverage polygon features using the Create New Feature task.

When you create new polygon features, arc features and a label point feature will be added to the coverage automatically.

If the polygon you create overlaps an existing polygon, the polygon feature and all arcs underneath will be split where features overlap.

Tip

Using the Auto Complete Polygon task

In addition to using the Create New Feature task, you can use the Auto Complete Polygon task and digitize a line to create new features. To learn more about the Auto Complete Polygon task, see 'Creating a new polygon with shared parts' in Chapter 8.

1. Click the Current Task dropdown arrow and click Create New Feature.

2. Click the Target layer dropdown arrow and click a coverage polygon layer.

3. Click the tool palette dropdown arrow and click the Sketch tool.

4. Click the map to create a new polygon sketch.

5. Double-click to finish the sketch and create a new polygon feature.

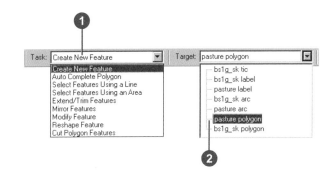

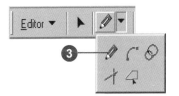

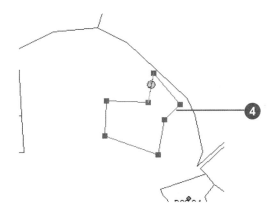

Creating region features

Region feature classes are *composite* feature classes. They are built from existing primitive features (polygons) that contain topological associations.

If you try to create a new region feature using the Create New Feature task, the sketch tools will be disabled.

Since a region feature is a composite of one or more polygon features, you need to select polygon features and use the Union command to create a new region.

Tip

Creating region feature classes

Region feature classes must be created using ArcToolbox prior to editing in ArcMap. To learn how to create region subclasses, see Using ArcToolbox.

Tip

Using the selection environment

It is often difficult to select polygon features that already form part of a region feature. You can use the Set Selectable Layers dialog box to make selections easier. For more information about selections, refer to Using ArcMap.

1. Click the Edit tool.

2. Select the polygon feature that you want to make a region out of.

3. Click the Target layer dropdown arrow and click the region subclass.

4. Click Editor and click Union from the Editor pulldown menu.

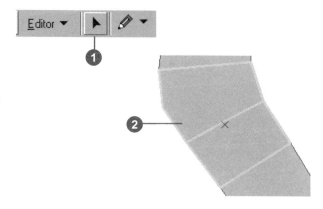

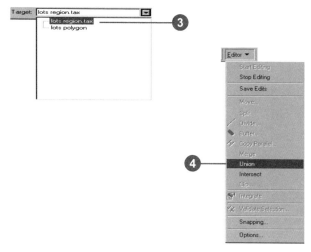

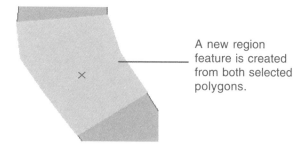

A new region feature is created from both selected polygons.

Creating route features

Route feature classes are composite feature classes composed of arc and section features with an associated measurement system.

Though section feature classes are not recognized by ArcMap, when you create a new route feature from selected arc features, section features will be created automatically.

1. Click the Edit tool.
2. Select the arc features out of which you want to make a route.
3. Click the Target layer dropdown arrow and click the route subclass.
4. Click Editor and click Union.

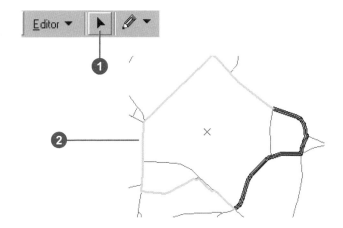

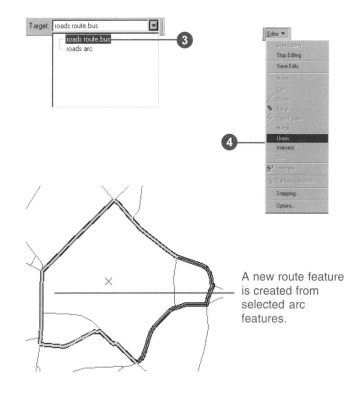

A new route feature is created from selected arc features.

Editing coverage features

The topology of your coverage determines the way in which you edit it.

If your coverage contains polygon and/or node features, then you need to use the topology tools to modify the shape of features. ArcMap will ensure the topological integrity of your coverage by disabling tools and commands that could break topological associations.

The Shared Edit tool lets you select the boundaries shared between features in a coverage. Editing shared boundaries ensures that the topology of your coverage is maintained. For example, when you select the boundary shared between two polygon features and reshape it using the Reshape task, both polygon features and all associated arc features are updated at the same time.

For more information about using the topology tools, see Chapter 8, 'Editing topological features'.

Tip

Creating a new arc feature
Another way that you can split polygons is by creating a new arc feature using the Create New Feature task.

Splitting a polygon

1. Click the Current Task dropdown arrow and click Auto Complete Polygon.

2. Click the Target layer dropdown arrow and click a coverage polygon layer.

3. Click the Sketch tool.

4. Create a sketch that starts and stops at any of the existing polygon boundaries to enclose the new polygon you are creating.

5. Double-click to finish the sketch.

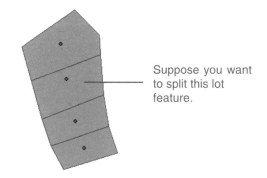

Suppose you want to split this lot feature.

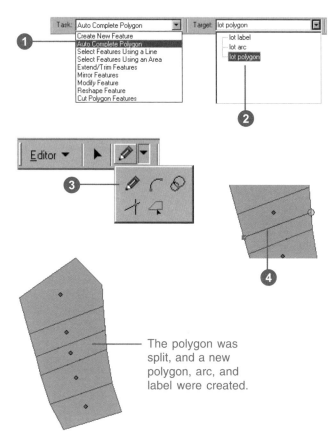

The polygon was split, and a new polygon, arc, and label were created.

Merging adjacent polygons

Polygon features share topological associations with arc features. Arc features form the perimeter of polygon features and the border between adjacent polygons. If you want to merge two adjacent polygons, you need to remove the arc features that form the border between them using the Shared Edit tool.

When you delete the arc features, the merged polygon will contain the attributes of the polygon with the lowest FID. Note that the FID is different than the internal record number <coverage>#. The label point of the old polygon will not be deleted. However, it will contain the attributes of the new polygon. To remove label points, select them using the Edit tool and click the Delete button.

1. Click the Shared Edit tool.
2. Select the boundary between the polygons you want to merge.
3. Click the Delete button.

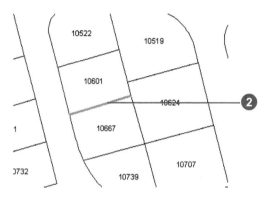

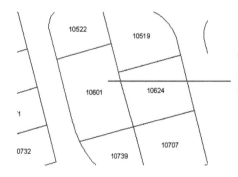

The border arc is removed, and the adjacent polygons merged.

Deleting a polygon

When your ArcMap selection contains topo features, editing tools and commands that could break the topological associations that feature contains are disabled.

For example, if you select a polygon feature using the Edit tool, the Delete command is disabled. Deleting the polygon would break the association that polygon has with the arc features that form its boundary and the label feature that stores attribute information.

You can delete topo features by selecting their shared boundaries using the Shared Edit tool. When you delete the shared boundary of a polygon feature, the arcs and label are deleted as well.

You also need to use the Shared Edit tool to delete arcs from a coverage that contains node features. If the arc feature you delete contains associations to node features that are floating, the node features will be deleted.

1. Click the Shared Edit tool.
2. Select a boundary of the polygon you want to delete.
3. Click the Delete command.

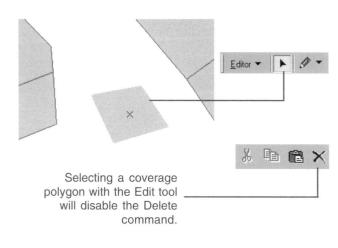

Selecting a coverage polygon with the Edit tool will disable the Delete command.

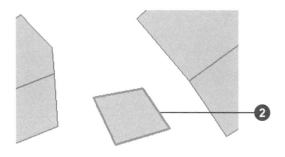

Moving a polygon

Coverage polygon features share topological associations with adjacent polygon features as well as arcs and labels. Each polygon feature is constrained so that it cannot overlap any other polygon feature in the same coverage.

Given this constraint, the Move command is disabled. The added functionality of the Edit tool, which lets you drag selected features, is also disabled. If you want to move a polygon feature, you must move all features that share topological associations with the polygon feature by using the Shared Edit tool.

1. Click the Shared Edit tool.

2. Select the boundary you wish to move.

3. Hold down the left mouse button and drag the boundary to a new location.

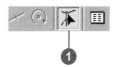

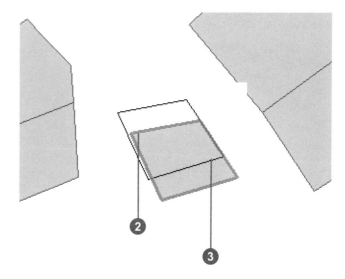

Splitting an arc

Arc features share topological associations with node and polygon features. If your arc coverage contains node features, splitting an arc feature will create a new node feature.

You can split an arc feature using the Split tool or by creating a new node feature using the Create New Feature task.

1. Click the Edit tool.
2. Select the arc you wish to split.
3. Click the split command.
4. Click where you want to split the arc.

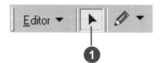

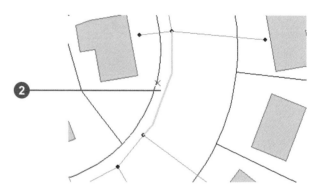

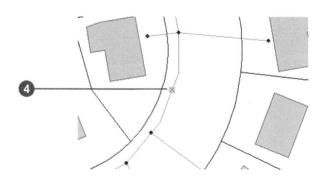

Merging connected arcs

Arc features that are connected to each other and have topology contain node features. A node feature is placed at the endpoints of each arc feature.

In order to merge two connected arc features together, you must delete the node that connects them. You need to select that node feature using the Shared Edit tool to delete it.

If the node feature that you delete is only connected to one arc feature, then deleting the node feature will delete the associated arc feature as well.

1. Click the Shared Edit tool.

2. Select the node feature that connects the two arc features.

3. Click the Delete command.

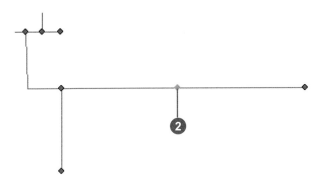

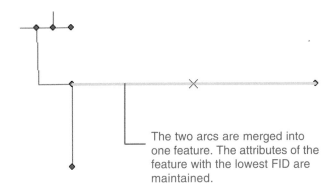

The two arcs are merged into one feature. The attributes of the feature with the lowest FID are maintained.

Deleting an arc

Arc features that contain topological associations with node and/or polygon features can only be deleted if you select them using the Shared Edit tool.

The Shared Edit tool discovers the topological association between arc features and node and/or polygon features.

If you delete an arc feature that borders two polygon features, then the polygon features will be merged together. For more information, see 'Merging adjacent polygons' in this appendix.

If you delete an arc feature that is either not connected to any other arc feature or is connected to only one arc feature, then all node features that would otherwise be left floating are removed.

1. Click the Shared Edit tool.
2. Select the arc feature you want to delete.
3. Click the Delete command.

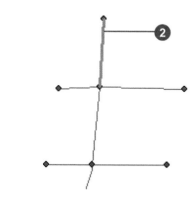

The selected arc feature and related node feature are deleted.

Moving an arc feature

If you want to move an arc feature that exists in a coverage containing either node or polygon feature classes, then you need to use the Shared Edit tool.

When you move an arc feature using the Shared Edit tool, the topological associations between arcs and nodes and/or polygons and arcs are maintained.

Tip

Extending an arc feature
If you select the endpoint of an arc feature using the Shared Edit tool, you can extend its length by dragging the endpoint.

1. Click the Shared Edit tool.
2. Select the arc you wish to move.
3. Hold down the left mouse button and drag the arc to a new location.

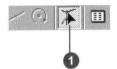

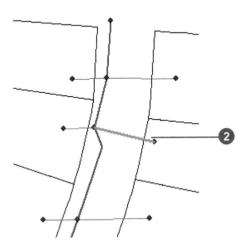

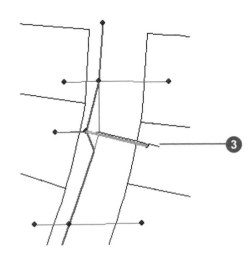

Merging region features

You can merge region features from the same coverage into one feature. The features must be a part of the same region feature class. You could use the Merge command to combine two lots based on ownership. If the selected lot features are not adjacent to each other, the merged feature will contain a multipart shape.

When you merge features, the original features are removed and the new feature's attributes are copied from the feature with the lowest ID number (the oldest feature).

1. Click the Edit tool.
2. Select the region features that you want to merge.
3. Click Editor and click Merge.

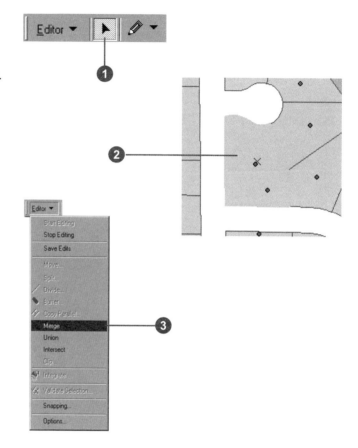

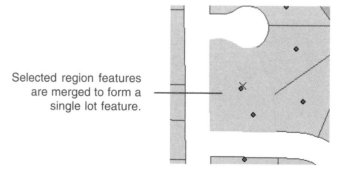

Selected region features are merged to form a single lot feature.

Merging route features

You can merge route features from the same coverage into one feature. The features must be a part of the same route feature class.

When you merge features, the original features are removed and the new feature's attributes are copied from the feature with the lowest ID number (the oldest feature).

1. Click the Edit tool.

2. Select the route features that you want to merge.

3. Click Editor and click Merge from the Editor pulldown menu.

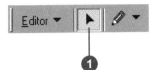

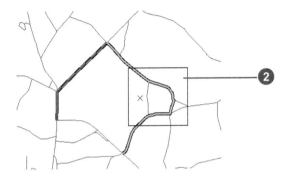

Glossary

active data frame

The data frame in a map that is currently being worked on—for example, the data frame to which layers are being added. The active data frame is shown in bold text in the ArcMap table of contents.

annotation

1. Descriptive text used to label features. It is used for display, not for analysis.

2. A feature class used to label other features. Information stored for annotation includes a text string, the location at which it is displayed, and a text symbol (color, font, size, and so on) for display.

arc–node topology

Arcs represent linear features and the borders of area features in a coverage. Every arc has a from-node, which is the first vertex in the arc, and a to-node, which is the last vertex. Nodes indicate the endpoints and intersections of arcs. They do not exist as independent features. Together they define the direction of the arc. Arc–node topology defines connectivity in coverages—arcs are connected to each other if they share a common node.

ArcInfo workspace

A file-based collection of coverages, grids, TINs, or shapefiles stored as a directory of folders in the file system.

ArcSDE

A gateway to a multiuser commercial RDBMS—for example, Oracle®, Microsoft® SQL Server™, Informix®, and DB2®. ArcSDE is an open, high-performance spatial data server that employs client/server architecture to perform efficient spatial operations and manage large, shared geographic data. Was known as SDE before 1999.

attribute

1. A piece of information describing a map feature. The attributes of a ZIP Code, for example, might include its area, population, and average per capita income.

2. A characteristic of a geographic feature described by numbers, characters, images, and CAD drawings, typically stored in tabular format and linked to the feature by a user-assigned identifier. For example, the attributes of a well might include depth and gallons per minute.

3. A column in a database table.

attribute domain

A named constraint in the database. An attribute constraint can be applied to a field of a subtype of a feature class or object class to make an attribute rule. Types of attribute domains include range and coded-value domains.

attribute table

A DBMS or other tabular file containing rows and columns. In ArcInfo, attribute tables are associated with a class of geographic features such as wells or roads. Each row represents a geographic feature. Each column represents one attribute of a feature, with the same column representing the same attribute in each row. See also feature attribute table.

Attributes dialog box

A dialog box that lets you view and edit attributes of features you've selected in ArcMap.

azimuth

An angle measured from north. Often used to define an oblique cylindrical projection or the angle of a geodesic between two points.

buffer

A zone of a specified distance around features. Both constant- and variable-width buffers can be generated on each feature's attribute values. The resulting buffer zones form polygons that are either inside or outside the specified buffer distance from each feature. Buffers are useful for proximity analysis (for example, find all stream segments within 300 feet of a proposed logging area).

CAD

See computer-aided design.

circle

A geometric shape for which the distance from the center to any point on the edge is equal.

cluster tolerance

The distance range in which all vertices and boundaries in a shapefile or feature dataset are considered identical, or coincident. A user-specified tolerance for the Integrate command in ArcMap. For example, if the cluster tolerance is set to 10 map units, after running Integrate there will be no more than one vertex within 10 map units of another.

coincident

Vertices or boundaries are coincident when they are within the cluster tolerance of one another. See also cluster tolerance.

column

The vertical dimension of a table. A column has a name and a data type applied to all values in the column. See also item, field, and attribute.

compress

The process of shrinking the size of a database or file. Improves performance by removing redundant information.

computer-aided design

An automated system for the design, drafting, and display of graphically oriented information.

conflict

In the versioning reconciliation process, if the same feature in the edit version and reconciliation version has been edited, the feature is said to be in conflict. Resolving the conflict requires you to make the decision as to the feature's correct representation using the Conflict Resolution dialog box.

connectivity

1. In a geodatabase, the state of edges and junctions in a logical network that controls flow, tracing, and pathfinding.

2. The topological identification in a coverage of connected arcs by recording the from- and to-node for each arc. Arcs that share a common node are connected. See also arc–node topology.

constraints

In real-world databases, an object's attributes can't have any particular value based solely on what data types and ranges a particular field type in the database allows. In reality, the permissible values are a range or list of values.

contiguity

In coverages, the topological identification of adjacent polygons by recording the left and right polygons of each arc. See also polygon–arc topology.

control points

Points you establish on a paper map to represent known ground points or specific locations. Control points are used to register a paper map before you begin digitizing features on it with a digitizer.

coordinate

A set of numbers that designate location in a given reference system such as x,y in a planar coordinate system or x,y,z in a three-dimensional coordinate system. Coordinates represent locations on the earth's surface relative to other locations.

coordinate system

1. A reference system used to measure horizontal and vertical distances on a planimetric map. A coordinate system is usually defined by a map projection; a spheroid of reference; a datum; one or more standard parallels; a central meridian; and possible shifts in the x- and y-directions to locate x,y positions of point, line, and area features.

2. In ArcInfo, a system with units and characteristics defined by a map projection. A common coordinate system is used to spatially register geographic data for the same area.

coverage

A file-based vector data storage format for storing the location, shape, and attributes of geographic features. A coverage usually represents a single theme, such as soils, streams, roads, or land use. It is one of the primary vector data storage formats for ArcInfo.

A coverage stores geographic features as primary features (such as arcs, nodes, polygons, and label points) and secondary features (such as tics, map extent, links, and annotation). Associated feature attribute tables describe and store attributes of the geographic features.

current task

During editing in ArcMap, a setting in the Current Task dropdown list that determines with which task the sketch construction tools (Sketch, Arc, Distance–Distance, and Intersection) will work. The current task is set by clicking a task in the Current Task dropdown list. All tasks in the Current Task dropdown list work with a sketch that you create. For example, the Create New Feature task uses a sketch you create to make a new feature. The Extend/Trim Feature task uses a sketch you create to determine where the selected feature will be extended or trimmed. The Cut Polygon Feature task uses a sketch you create to determine where the polygon will be cut.

dangle tolerance

The minimum length allowed for dangling arcs in coverages in the ArcInfo Clean process. Clean removes dangling arcs that are shorter than the dangle length. Also known as the dangle length.

dangling arc

In coverages, an arc having the same polygon on both its left and right sides and having at least one node that does not connect to any other arc. It often identifies where a polygon does not close properly (for example, undershoot), where arcs don't connect properly, or where an arc was digitized past its intersection with another arc (for example, overshoot). A dangling arc is not always an error. For example, dangling arcs can represent cul-de-sacs in street centerline maps.

data frame

In ArcMap, a frame on the map that displays layers occupying the same geographic area. You may have one or more data frames on your map depending upon how you want to organize your data. For instance, one data frame might highlight a study area, and another might provide an overview of where the study area is located.

data integrity

Maintenance of data values according to data model and data type. For example, to maintain integrity, numeric columns will not accept character data.

data source

Any geographic data, such as a coverage, shapefile, raster, or feature class, in a geodatabase.

data type

The characteristic of columns and variables that defines what types of data values they can store. Examples include character, floating point, and integer.

data view

An all-purpose view in ArcMap for exploring, displaying, and querying geographic data. This view hides all map elements such as titles, North arrows, and scale bars. See also layout view.

dataset

1. Any feature class, table, or collection of feature classes or tables in the geodatabase.

2. A named collection of logically related data items arranged in a prescribed manner.

decimal degrees

Degrees of latitude and longitude expressed as a decimal rather than in degrees, minutes, and seconds.

digitizing

1. To encode geographic features in digital form as x,y coordinates.

2. The process of converting the features on a paper map into digital format. When you digitize a map, you use a digitizing tablet, or digitizer, connected to your computer and trace over features with a digitizer puck, which is similar to a mouse. The x,y coordinates of these features are automatically recorded and stored as spatial data.

digitizing mode

Also called absolute mode, digitizing mode is one of the ways in which a digitizing tablet operates. In digitizing mode, the location of the tablet is mapped to a specific location on the screen. Moving the digitizer puck on the tablet surface causes the screen pointer to move to precisely the same position.

distance units

The units—for example, feet, miles, meters, or kilometers—ArcMap uses to report measurements, dimensions of shapes, and distance tolerances and offsets.

double precision

Refers to a high level of coordinate accuracy based on the possible number of significant digits that can be stored for each coordinate. ArcInfo datasets can be stored in either single- or double-precision coordinates. Double-precision coverages store up to 15 significant digits per coordinate (typically, 13 to 14 significant digits), retaining the accuracy of much less than 1 meter at a global extent. See also single precision.

edit cache

A setting in ArcMap that causes the features visible in the current map extent to be held in memory on your local machine. Designed to be used when working with large amounts of data, an edit cache results in faster editing because ArcMap doesn't have to retrieve the data from the server.

edit session

In ArcMap, all editing takes place within an edit session. An edit session begins when you choose Start Editing from the Editor menu and ends when you choose Stop Editing.

Editor toolbar

A toolbar that lets you create and modify features and their attributes in ArcMap.

ellipse

A geometric shape equivalent to a circle that is viewed obliquely; a flattened circle.

extent

The coordinates defining the minimum bounding rectangle (that is, xmin, ymin and xmax, ymax) of a data source. All coordinates for the data source fall within this boundary.

feature

1. An object class in a geodatabase that has a field of type geometry. Features are stored in feature classes.

2. A representation of a real-world object.

3. A point, line, or polygon in a coverage or shapefile.

feature attribute table

A table used to store attribute information for a specific coverage feature class. ArcInfo maintains the first several items of these tables. Feature attribute tables supported for coverages include the following:

<cover>.PAT	for polygons or points
<cover>.AAT	for arcs
<cover>.NAT	for nodes
<cover>.RAT	for routes
<cover>.SEC	for sections
<cover>.PAT	for regions
<cover>.TAT	for annotation (text)

where <cover> is the coverage name.

feature class

1. The conceptual representation of a geographic feature. When referring to geographic features, feature classes include point, line, area, and annotation. In a geodatabase, an object class that stores features and has a field of type geometry.

2. A classification describing the format of geographic features and supporting data in a coverage. Coverage feature classes for representing geographic features include point, arc, node, route-system, route, section, polygon, and region. One or more

coverage features are used to model geographic features; for example, arcs and nodes can be used to model linear features such as street centerlines. The tic, annotation, link, and boundary feature classes provide supporting data for coverage data management and viewing.

3. The collection of all the point, line, or polygon features or annotation in a CAD dataset.

feature dataset

In geodatabases, a collection of feature classes that share the same spatial reference. Because the feature classes share the same spatial reference, they can participate in topological relationships with each other such as in a geometric network. Several feature classes with the same geometry may be stored in the same feature dataset. Object classes and relationship classes can also be stored in a feature dataset.

field

A column in a table. Each field contains the values for a single attribute.

fuzzy tolerance

An extremely small distance used to resolve inexact intersection locations due to the limited arithmetic precision of computers. It defines the resolution of a coverage resulting from the Clean operation or a topological overlay operation such as Union, Intersect, or Clip.

geodatabase

An object-oriented geographic database that provides services for managing geographic data. These services include validation rules, relationships, and topological associations. A geodatabase contains feature datasets and is hosted inside of a relational database management system.

geodatabase data model

Geographic data model that represents geographic features as objects in an object-relational database. Features are stored as rows in a table; geometry is stored in a shape field. Supports sophisticated modeling of real-world features. Objects may have custom behavior.

georelational data model

A geographic data model that represents geographic features as an interrelated set of spatial and descriptive data. The georelational model is the fundamental data model used in coverages.

index

A special data structure used in a database to speed searching for records in tables or spatial features in geographic datasets. ArcInfo supports both spatial and attribute indexes.

intersect

The topological integration of two spatial datasets that preserves features that fall within the area common to both input datasets. See also union.

item

1. A column of information in an INFO table.

2. An element in the Catalog tree. The Catalog tree can contain both geographic data sources and nongeographic elements such as folders, folder connections, and file types.

label point

A feature class in a coverage used to represent point features and identify polygons.

layer

1. A collection of similar geographic features—such as rivers, lakes, counties, or cities—of a particular area or place for display on a map. A layer references geographic data stored in a data source, such as a coverage, and defines how to display it. You can create and manage layers as you would any other type of data in your database.

2. A feature class in a shared geodatabase managed with SDE 3.

layout view

The view for laying out your map in ArcMap. Layout view shows the virtual page upon which you place and arrange geographic data and map elements—such as titles, legends, and scale bars—for printing. See also data view.

left–right topology

The topological data structure ArcInfo uses to represent contiguity between polygons. Left–right topology supports analysis functions such as adjacency. See also topology.

map

1. A graphical presentation of geographic information. It contains geographic data and other elements, such as a title, North arrow, legend, and scale bar. You can interactively display and query the geographic data on a map and also prepare a printable map by arranging the map elements around the data in a visually pleasing manner.

2. The document used in ArcMap that lets you display and work with geographic data. A map contains one or more layers of geographic data and various supporting map elements such as scale bars. Layers on a map are contained in data frames. A data frame has properties such as scale, projection, and extent and also graphic properties such as where it is located on your map's page. Some maps have one data frame, while other more advanced maps may have several data frames.

map document

In ArcMap, the disk-based representation of a map. Map documents can be printed or embedded into other documents. Map documents have a .mxd file extension.

map units

The units—for example, feet, miles, meters, or kilometers—in which the coordinates of spatial data are stored.

merge policy

In geodatabases, all attribute domains have a merge policy associated with them. When two features are merged into a single feature in ArcMap, the merge policies dictate what happens to the value of the attribute to which the domain is associated. Standard merge policies are default value, sum, and weighted average.

minimum bounding rectangle

A rectangle, oriented to the x- and y-axes, that bounds a geographic feature or a geographic dataset. It is specified by two coordinates: xmin, ymin and xmax, ymax. For example, the extent defines a minimum bounding rectangle for a coverage.

multipart feature

A feature that is composed of more than one physical part but only references one set of attributes in the database. For example, in a layer of states, the State of Hawaii could be considered a multipart feature. Although composed of many islands, it would be recorded in the database as one feature.

multipoint feature

A feature that consists of more than one point but only references one set of attributes in the database. For example, a system of oil wells might be considered a multipoint feature, as there is a single set of attributes for the main well and multiple well holes.

multiuser geodatabase

A geodatabase in an RDBMS served to client applications (for example, ArcMap) by ArcSDE. Multiuser geodatabases can be very large and support multiple concurrent editors. Supported on a variety of commercial RDBMSs including Oracle, Microsoft SQL Server, IBM DB2, and Informix.

null value

The absence of a value. A geographic feature for which there is no associated attribute information.

overshoot

That portion of an arc digitized past its intersection with another arc. See also dangling arc.

pan

To move the viewing window up, down, or sideways to display areas in a geographic dataset that, at the current viewing scale, lies outside the viewing window.

parametric curve

A curved segment that has only two vertices as endpoints, instead of being made of numerous vertices. You can create a parametric curve using the Arc tool or the Tangent Curve command in the ArcMap Editor. Also known as a true curve.

personal geodatabase

A geodatabase, usually on the same network as the client application (for example, ArcMap), that supports one editor at a time. Personal geodatabases are managed in a Microsoft Jet Engine database.

planar topology

Represent collections of topological feature classes that share geometry among their boundaries. One or more line and polygon feature classes that share geometry participate in a common planar topology. Updating shared boundaries updates all features in the topology.

point

A single x,y coordinate that represents a single geographic feature such as a telephone pole.

point mode digitizing

One of two methods of digitizing features using the ArcMap Editor's Sketch tool or from a paper map using a digitizer. With point mode digitizing, you can create or edit features by digitizing a series of precise points, or vertices. Point mode digitizing is effective when precise digitizing is required—for example, when digitizing a perfectly straight line. See also stream mode digitizing.

polygon

A two-dimensional feature representing an area such as a state or county.

polygon–arc topology

PAT. A coverage polygon is made up of arcs that define the boundary and a label point that links the polygon feature to an attribute record in the coverage PAT. ArcInfo stores polygons topologically as a list of arcs and a label that make up each polygon.

precision

Refers to the number of significant digits used to store numbers and, in particular, coordinate values. Precision is important for accurate feature representation, analysis, and mapping. ArcInfo supports single and double precision.

preliminary topology

In coverages, refers to incomplete region or polygon topology. Region topology defines region–arc and region–polygon relationships. A topological region has both the region–arc relationship and the region–polygon relationship. A preliminary region has the region–arc relationship but not the region–polygon relationship. In other words, preliminary regions have no polygon topology. Polygon topology defines polygon–arc-label point relationships. A preliminary polygon has the polygon–label point relationship but not the polygon–arc relationship. Coverages with preliminary topology have red in their icons in the Catalog.

projection

A mathematical formula that transforms feature locations from the earth's curved surface to a map's flat surface. A projected coordinate system employs a projection to transform locations expressed as latitude and longitude values to x,y coordinates. Projections cause distortions in one or more of these spatial properties: distance, area, shape, and direction.

property

An attribute of an object defining one of its characteristics or an aspect of its behavior. For example, the Visible property affects whether a control can be seen at run time. You can set a data source's properties using its Properties dialog box.

query

A question or request used for selecting features. A query often appears in the form of a statement or logical expression. In ArcMap, a query contains a field, an operator, and a value.

radius

The distance from the center to the outer edge of a circle or circular curve.

RDBMS

Relational database management system. A database management system with the ability to access data organized in tabular files that can be related to each other by a common field (item). An RDBMS has the capability to recombine the data items from different files, providing powerful tools for data usage. ArcSDE supports several commercial RDBMSs.

record

1. In an attribute table, a single "row" of thematic descriptors. In SQL terms, a record is analogous to a tuple.

2. A logical unit of data in a file. For example, there is one record in the ARC file for each arc in a coverage.

relate

An operation that establishes a temporary connection between corresponding records in two tables using an item common to both (for example, key attributes). Each record in one table is connected to those records in the other table that share the same value for the common item. See also relational join.

relational join

The operation of relating and physically merging two attribute tables using their common item.

relationship

An association or link between two objects in a database. Relationships can exist between spatial objects (features in feature classes) or nonspatial objects (rows in a table), or between spatial and nonspatial objects.

row

1. A record in an attribute table. The horizontal dimension of a table composed of a set of columns containing one data item each.

2. A horizontal group of cells in a raster.

scanning

The process of capturing data in raster format with a device called a scanner. Some scanners also use software to convert raster data to vector data.

segment

A line that connects vertices. For example, in a sketch of a building, a segment would represent one wall.

select

To choose from a number or group of features or records; to create a separate set or subset.

Selectable Layers list

A list on the Selection menu in ArcMap that lets you choose from which layers you can select. For example, suppose you wanted to select a large number of buildings by drawing a box around them but selected a parcel by mistake as you drew the selection box. To avoid this, you might uncheck the Parcels layer in the Selectable Layers list so that parcels cannot be selected.

selected set

A subset of the features in a layer or records in a table. ArcMap provides several ways to select features and records graphically or according to their attribute values.

selection anchor

When editing in ArcMap, a small "x" located in the center of selected features. The selection anchor is used when you move features using snapping. It is the point on the feature or group of features that will be snapped to the snapping location. This is also the point around which your selection will rotate when you use the Rotate tool and around which your feature will scale when you use the Scale tool.

shape

The characteristic appearance or visible form of a geographic object. Geographic objects can be represented on a map using one of three basic shapes: points, lines, or polygons.

shapefile

A vector data storage format for storing the location, shape, and attributes of geographic features. A shapefile is stored in a set of related files and contains one feature class.

shared boundary

A segment or boundary common to two features. For example, in a parcel database, adjacent parcels will share a boundary. Another example might be a parcel that shares a boundary on one side with a river. The segment of the river that coincides with the parcel boundary would share the same coordinates as the parcel boundary.

shared vertex

A vertex common to multiple features. For example, in a parcel database, adjacent parcels will share a vertex at the common corner.

single precision

Refers to a level of coordinate accuracy based on the number of significant digits that can be stored for each coordinate. Single-precision numbers store up to seven significant digits for each coordinate, retaining a precision of ±5 meters in an extent of 1,000,000 meters. ArcInfo datasets can be stored as either single- or double-precision coordinates. See also double precision.

sketch

When editing in ArcMap, a shape that represents a feature's geometry. Every existing feature on a map has an alternate form, a sketch. A sketch lets you see exactly how a feature is composed, with all vertices and segments of the feature visible. To modify a feature, you must modify its sketch. To create a feature, you must first create a sketch. You can only create line and polygon sketches, as points have neither vertices nor segments.

Sketches help complete the current task. For example, the Create New Feature task uses a sketch you create to make a new feature. The Extend/Trim Feature task uses a sketch you create to determine where the selected feature will be extended or trimmed. The Cut Polygon Feature task uses a sketch you create to determine where the polygon will be cut into two features.

sketch constraints

In ArcMap editing, the angle or length limitations you can place on segments you're creating. These commands are available on the Sketch tool context menu. For example, you can set a length constraint that specifies that the length of the segment you're creating will be 50 map units. At whatever angle you create that segment, its length will be constrained to 50 map units.

Angle constraints work in the same way. For example, you can set an angle constraint that specifies that the angle of the segment you're creating will be 45 degrees measured from another feature that already exists. At whatever length you create that segment, its angle will be constrained to 45 degrees.

sketch operations

In ArcMap, editing operations that are performed on an existing sketch. Examples are Insert Vertex, Delete Vertex, Flip, Trim, Delete Sketch, Finish Sketch, and Finish Part. All of these operations are available from the Sketch context menu, which is available when you right-click on any part of a sketch using any editing tools.

snapping

The process of moving a feature to coincide exactly with coordinates of another feature within a specified snapping distance or tolerance.

snapping environment

Settings in the ArcMap Editor's Snapping Environment window and Editing Options dialog box that help you establish exact locations in relation to other features. You determine the snapping environment by setting a snapping tolerance, snapping properties, and a snapping priority.

snapping priority

During ArcMap editing, the order in which snapping will occur by layer. You can set the snapping priority by dragging the layer names in the Snapping Environment window to new locations.

snapping properties

In ArcMap editing, a combination of a shape to snap to and a method for what part of the shape you will snap to. You can set your snapping properties to have a feature snap to a vertex, edge, or endpoint of features in a specific layer. For example, a layer snapping property might let you snap to the vertices of buildings. A more generic, sketch-specific snapping property might let you snap to the vertices of a sketch you're creating.

snapping tolerance

During ArcMap editing, the distance within which the pointer or a feature will snap to another location. If the location being snapped to (vertex, boundary, midpoint, or connection) is within the distance you set, the pointer will automatically snap. For example, if you want to snap a power line to a utility pole and the snapping tolerance is set to 25 pixels, whenever the power line comes within a 25-pixel range of the pole it will automatically snap to it. Snapping tolerance can be measured using either map units or pixels.

spatial join

A type of spatial analysis in which the attributes of features in two different layers are joined together based on the relative locations of the features.

spatial reference

Describes both the projection and spatial domain extent for a feature dataset or feature class in a geodatabase.

SQL

Structured Query Language. A syntax for defining and manipulating data from a relational database. Developed by IBM® in the 1970s, it has become an industry standard for query languages in most relational database management systems.

stream mode digitizing

One of the two methods of digitizing features from a paper map. Also known as streaming, stream mode digitizing provides an easy way to capture features when you don't require much precision—for example, to digitize rivers, streams, and contour lines. With stream mode, you create the first vertex of the feature and trace over the rest of the feature with the digitizer puck. You can also use digitize in stream mode with the ArcMap Sketch tool when editing "freehand". See also point mode digitizing.

stream tolerance

The interval at which vertices are added along the feature you're digitizing in stream mode. When streaming, vertices are automatically created at a defined interval as you move the mouse. For example, if the stream tolerance is set to 10 map units, you must move the pointer at least 10 map units before the next vertex will be created. If you move the pointer more than 10 map units, there may be more space between vertices, but there will always be a minimum interval of 10 map units. Stream tolerance is measured in map units. See also stream mode digitizing.

symbol

A graphic pattern used to represent a feature. For example, line symbols represent arc features; marker symbols, points; shade symbols, polygons; and text symbols, annotation. Many characteristics define symbols including color, size, angle, and pattern.

symbology

The criteria used to determine symbols for the features in a layer. A characteristic of a feature may influence the size, color, and shape of the symbol used.

table

Information formatted in rows and columns. A set of data elements that has a horizontal dimension (rows) and a vertical dimension (columns) in an RDBMS. A table has a specified number of columns but can have any number of rows. See also attribute table.

table of contents

In ArcMap, lists all the data frames and layers on the map and shows what the features in each layer represent.

tabular data

Descriptive information that is stored in rows and columns and can be linked to map features.

target layer

Used in ArcMap editing, a setting in the Target layer dropdown list that determines to which layer new features will be added. The target layer is set by clicking a layer in the Target layer dropdown list. For instance, if you set the target layer to Buildings, any features you create will be part of the Buildings layer. You must set the target layer whenever you're creating new features— whether you're creating them with the Sketch tool, by copying and pasting, or by buffering another feature.

tic

Registration or geographic control points for a coverage representing known locations on the earth's surface. Tics allow all coverage features to be recorded in a common coordinate system such as Universal Transverse Mercator (UTM). Tics are used to register map sheets when they are mounted on a digitizer and to transform the coordinates of a coverage, for example, from digitizer units (inches) to the appropriate values for a coordinate system (which are measured in meters for UTM).

tolerances

A coverage uses many processing tolerances (fuzzy, tic match, dangle length) and editing tolerances (weed, grain, edit distance, snap distance, and nodesnap distance). Stored in a TOL file, ArcInfo uses the values as defaults in many automation, editing, and processing operations. You can edit a coverage's tolerances using its Properties dialog box in ArcCatalog.

topological association

The spatial relationship between features that share geometry such as boundaries and vertices. When you edit a boundary or vertex shared by two or more features using the topology tools in the ArcMap Editor, the shape of each of those features is updated.

topological feature

A feature that supports network connectivity that is established and maintained based on geometric coincidence.

topology

1. In geodatabases, relationships between connected features in a geometric network or shared borders between features in a planar topology.

2. In coverages, the spatial relationships between connecting or adjacent features (for example, arcs, nodes, polygons, and points). The topology of an arc includes its from- and to-nodes and its left and right polygons. Topological relationships are built from simple elements into complex elements: points (simplest elements), arcs (sets of connected points), areas (sets of connected arcs), and routes (sets of sections, which are arcs or portions of arcs). Redundant data (coordinates) is eliminated because an arc may represent a linear feature, part of the boundary of an area feature, or both.

transaction

A logical unit of work as defined by a user. Transactions can be data definition (create an object), data manipulation (update an object), or data read (select from an object).

true curve

See parametric curve.

undershoot

An arc that does not extend far enough to intersect another arc. See also dangling arc.

union

A topological overlay of two polygonal spatial datasets that preserves features that fall within the spatial extent of either input dataset; that is, all features from both coverages are retained. See also intersect.

validation rule

Validation rules can be applied to objects in the geodatabase to ensure that their state is consistent with the system that the database is modeling. The geodatabase supports attribute, connectivity, relationship, and custom validation rules.

version

In geodatabases, an alternative representation of the database that has an owner, a description, a permission (private, protected, or public), and a parent version. Versions are not affected by changes occurring in other versions of the database.

vertex

One of a set of ordered x,y coordinates that defines a line or polygon feature.

virtual page

The map page, as seen in layout view.

wizard

A tool that leads a user step by step through an unusually long, difficult, or complex task.

workspace

A container for file-based geographic data. This can be a folder that contains shapefiles, an ArcInfo workspace that contains coverages, a personal geodatabase, or an ArcSDE database connection.

Index

Coordinate
  defined 211
Coordinate system
  defined 211
Copying
  a line parallel to an existing line 124
  and pasting attributes 73
  and pasting features 73
Copying and pasting features 27
Coverage
  defined 211
Coverages
  composite feature classes 190
  creating new coverage features 192
  editing 53–54
  editing features 198
  primary feature classes 190
  secondary feature class 190
  simple features 191
  tolerances 192
  topo features 191
  topological associations 190
  topology 190
Creating a new arc feature 198
Current task
  defined 211

D

Dangle tolerance
  defined 212
Dangling arc
  defined 212
Data
  adding for editing 60
  loading from a geodatabase 60
  stopping the drawing of 60
Data frames
  defined 212
  editing a map with multiple 63

Data integrity
  defined 212
Data model. *See* Georelational data model
Data source
  defined 212
Data type
  defined 212
Data view
  defined 212
Dataset
  defined 212
Decimal degrees
  defined 212
Deflection 94–95
Deleting
  features 72
Digitizer
  aligning the map on 112
  attaching the map to 112
  configuring puck buttons
    for streaming 122
    using programming code 112, 119, 122
    using WinTab manager setup program 112
  creating features with 111
  installing driver software 112
  setting up 22, 112
Digitizing
  a projected map 112
  defined 212
  digitizer tab missing 114
  finishing your digitizing session 26
  freehand 111, 117
  in absolute mode 23
  in digitizing (absolute) mode 117, 118–119
  in mouse (relative) mode 117
  in point mode 24, 117, 118–119
  in stream mode 25, 117, 120–121, 122
  installing digitizer driver software 112
  preparing the map 22, 112
  registering your map 22

Digitizing (continued)
  switching between digitizing and mouse
    modes 117
  switching between point and stream modes
    121
  using snapping 119
Digitizing mode
  defined 212
Digitizing tablet. *See* Digitizer
Distance units
  defined 213
Distance-Distance tool. *See* Editing tools
Dividing a line feature. *See* Placing points along
    a line
Double precision
  defined 213
Dragging features 68. *See also* Moving features

E

Edit cache
  creating 61, 62
  defined 213
  toolbar 61
  zooming to extent 62
Edit session
  defined 213
  starting 61
  stopping 63
Editing
  a map with more than one data frame 63
  adding and deleting sketch vertices 152–153
  adding the Editor toolbar 59
  combining features from different layers 132
  copying a line parallel to an existing line 124
  copying and pasting features 73
  coverages 53–54
  creating a mirror image of a feature 128
  creating a segment that is a parametric curve
    102–103
  creating a sketch 56

Layers (continued)
  defined 215
  setting selectable 67
Layout view
  defined 215
  editing in 63
Left-right topology
  defined 215
Line features
  copying 124
  creating 17, 56, 76, 88–89, 90
  creating a new arc feature 192
  deleting coverage arcs 204
  extending 31, 145
  extending coverage arcs 205
  flipping 147
  merging connected arcs 203
  moving coverages arcs 205
  placing points along 148
  reshaping. *See* Reshaping features
  splitting 138–139
  splitting coverage arcs 202
  trimming 142–144
Loading
  data from a geodatabase 60
Loading data
  from a CAD drawing 42

M

M values
  editing 162
Map
  defined 215
Map document
  defined 215
Map units
  defined 215
Measurements
  adding vertices using 18

Measurements (continued)
  setting number of decimal places for reporting
    74
  viewing 74
Merge policy
  defined 215
Merging features 130
Minimum bounding rectangle
  defined 215
Mirror image 128
Missing Digitizer tab 114
Modifying features
  by moving vertices 154–155
  shortcut 142
Moving features
  by dragging 29, 68
  by rotating 69, 71
  relative to their current position 69
  undoing a move 68
  using delta x,y coordinates 68–69, 69–70
  using the snapping environment 69
Moving vertices. *See* Reshaping features
Multipart features
  creating 90–91, 132
  defined 215
  described 89
  merging features to create 130
  removing parts 161
  shortcut for finishing a part 90
Multipoint features
  creating 86–87
  defined 215
  described 84
Multiuser geodatabase
  defined 216

N

Node features
  creating 193

Null value
  defined 216

O

Overshoot
  defined 216

P

Pan
  defined 216
Parallel segments. *See* Segments: creating
    parallel to existing
Parametric curves
  creating 102–103
  defined 216
Pasting
  features. *See* Copying and pasting features
Perpendicular segments. *See* Segments: creating
    perpendicular to existing
Personal geodatabase
  defined 216
Placing points along a line 148
Planar topology
  defined 216
Point
  defined 216
Point features
  creating 56, 76, 80–85, 90, 194
Point mode digitizing. *See* Digitizing: in point
    mode
  defined 216
Polygon
  defined 216
Polygon features
  creating 12, 13, 56–57, 88–89, 179, 195
  cutting a polygon shape out of 141
  deleting 200
  merging 199
  moving 201

Polygon features (continued)
  reshaping. *See* Reshaping features
  splitting 138–139, 198
  squaring 92–93
Polygon-arc topology
  defined 216
Precision
  defined 216
Preliminary topology
  defined 217
Projection
  defined 217
Properties
  changing for a sketch 159
Property
  defined 217

Q

Query
  defined 217

R

Radius
  defined 217
RDBMS (relational database management
    system)
  defined 217
Record
  defined 217
Region features
  creating 196
  creating region feature classes 196
  merging 206
Registering a map
  adding additional control points 115
  described 113–114
  digitizing control points 113
  ensuring accuracy when 116

Registering a map (continued)
  enterring ground coordinates 113
  error reporting 112, 114
  establishing control points 112
  for the first time 113
  removing ground coordinate records 115
  saving ground coordinates 113–114
  using existing tic files or saved coordinates
      115–116
Relate
  defined 217
Relational join
  defined 217
Relationship
  defined 217
Relative mode. *See* Digitizing: in mouse
      (relative) mode
Reporting measurements. *See* Measurements:
      setting number of decimal places for
      reporting
Reshaping a shared boundary 37
Reshaping features
  by adding vertices 152–153
  by deleting vertices 152–153
  by moving vertices 154–155
  using a sketch you draw 150–151
RMS error 112, 114
Root mean square error. *See* RMS error
Rotating a point's symbology 71
Rotating features 28, 69, 71
Route features
  creating 197
  merging 207
Row
  defined 218

S

Saving edits 16, 46, 62
Scaling features 30, 163

Scanning
  defined 218
Segment Deflection. *See* Editing: creating
      segments: using angles from existing
      segments
Segments
  creating
    at an angle from the last segment 96–97
    parallel to existing 19, 99
    parametric curves 19, 102–103
    perpendicular to existing 99
    using angles and lengths 94–95
    using angles from existing segments 98
  defined 218
Select
  defined 218
Selectable layers list
  defined 218
Selected set
  defined 218
Selecting features
  described 64
  removing features from the selection 64
  setting selectable layers 67
  using a line 65
  using a polygon 66
  using the Edit tool 64
  using the Selection menu 66
Selection anchor
  and snapping 70
  defined 218
  described 65
  moving
    when rotating or snapping a feature 70
    when scaling a feature 163
Shape
  defined 218
Shapefile
  defined 218
Shared boundary
  defined 218

Tracing. *See* Editing tools
Transaction
    defined 221
Trimming line features 142–144
True curve. *See* Parametric curves
    defined 221

U

Undershoot
    defined 221
Union. *See* Combining features from different
        layers
    defined 222

V

Validation rule
    defined 222
VBA
    configuring digitizer puck buttons using
        112, 119
Vector datasets
    comparing the structure of 53
Version
    defined 222
Vertex
    defined 222
Vertices
    adding 57, 152, 160
    creating 80–85
        using absolute coordinates 20
    deleting 57, 88, 119, 152
    deleting multiple while streaming with a
        digitizer 121
    moving
        by dragging 154
        by specifying x,y coordinates 155–156
        relative to the current location 157–158
    undoing and redoing 84, 89

Virtual page
    defined 222
Visual Basic for Applications. *See* VBA

W

Wizard
    defined 222
Workspaces
    defined 222
    editing a map with multiple 61

Z

Z values
    editing 162
    using the current control 162